Palgrave Studies in Economic History

Series Editor
Kent Deng
London School of Economics
London, UK

Palgrave Studies in Economic History is designed to illuminate and enrich our understanding of economies and economic phenomena of the past. The series covers a vast range of topics including financial history, labour history, development economics, commercialisation, urbanisation, industrialisation, modernisation, globalisation, and changes in world economic orders.

More information about this series at
http://www.palgrave.com/gp/series/14632

Eric L. Jones

Landscape History and Rural Society in Southern England

An Economic and Environmental Perspective

palgrave
macmillan

Eric L. Jones
University of Buckingham
Buckingham, UK

ISSN 2662-6497 ISSN 2662-6500 (electronic)
Palgrave Studies in Economic History
ISBN 978-3-030-68615-4 ISBN 978-3-030-68616-1 (eBook)
https://doi.org/10.1007/978-3-030-68616-1

Cover credit: Chronicle/Alamy Stock Photo

This Palgrave Macmillan imprint is published by the registered company Springer Nature Switzerland AG
The registered company address is: Gewerbestrasse 11, 6330 Cham, Switzerland

For Patrick Dillon

Acknowledgements

I am grateful to John Anderson, Deborah Cassell, Patrick Dillon, Shaun Keneally and John Price for reading parts of the manuscript; to Syd Flatman for accompanying me on many site visits; and above all to my wife, Sylvia, and son, Christopher, for editing the text.

Contents

PART I

Preliminaries

CHAPTER 1

Strategic Locations

Economic history perches uneasily, but creatively, between the abstractions of economics and the conventions of history. Environmental history follows as an uneasy adjunct of economic and agricultural history, dealing with nature, resources and the landscape. What can features on the ground tell us about economic history and ecology, about how the subjects relate and how the human habitat changed? What were the feedback effects—how did economic change modify the environment and become shaped by it in turn? The localities described in this book exemplify land-use change and reclamation, communal land management, drove roads, arable colonisation of the downs, the layout of parkland and resources including fodder, wool and wood.[1] Most occupations have left marks and signs to be sought out and interpreted. There is no need to resort to the cliché that minor sites and fading marks contain the world in a grain of sand, all that need be said is that each exhumes half-forgotten activities. They are clues rather than microcosms. It is the task of the environmental historian to pursue them and examine their consequences.

The value of ground evidence is that it sometimes provides the keys to economic and ecological processes which texts do not reveal. When texts do supply hints they are usually to be found in quite fugitive writings. Beyond what is evident in non-standard sources, conjecture may be needed. The military scholar, Liddell Hart, called this surmising what is

E. L. Jones, *Landscape History and Rural Society in Southern England*, Palgrave Studies in Economic History, https://doi.org/10.1007/978-3-030-68616-1_1

'the other side of the hill'. We also need to do something recommended in urban history: stand and watch as a town changes around us in our mind's eye, pursuing the implications at the next stage of research. Without rounded descriptions the history of the environment may be distorted and will certainly be incomplete.

Studies based purely on texts and documents may include descriptions backed by particular examples, more verisimilitude than verity; the tabulation of numerical data, ingenious but merely assumed to be faithful proxies and definitely more abstract than readable; or the biographies of single parishes. None of these approaches is entirely compelling and although a one-place study is likely to be the most persuasive, even that has limitations. Gilbert White, whose opinion commands universal respect, did say that the place most examined produces the most results. While this is true in an absolute sense it scouts the law of diminishing returns. Only the rarest of the rare among parishes, such as Kibworth, Leicestershire, is ever likely to yield adequate evidence to follow its evolution throughout every period.[2] Kibworth's documentation is utterly exceptional and serves to emphasise that the records of most villages contain gaps which have to be filled by interpolating material from other places.

Strategic Locations

A one-place study must confront questions about representativeness. How is it known that a given parish shared the experience of its neighbours, let alone of the region or country at large? I have coped with the difficulty by studying smallish clusters of parishes centred on places across Southern England where I have lived. They are my 'strategic locations'. As a geographical device it is semi-arbitrary but is rescued by covering enough cases to offset peculiarities or lapses in any one. Each cluster has some ecological unity. Administrative boundaries may break this apart, as they split up the management of land belonging to families with portfolios of property in more than one county, but on the whole, each group of parishes falls into a recognisable ecosystem.

The localities considered here in detail are in Hampshire, West Berkshire, Oxfordshire and Gloucestershire, with side excursions to East Berkshire, Wiltshire and Devon. Besides growing up in Andover, Hampshire, I have lived in parishes elsewhere within reach of universities where I was working: Eynsham and Kingston Bagpuize near Oxford; Stratfield Mortimer, west of Reading, and Wantage, formerly in Berkshire, now in Oxfordshire, but brought within reach of Reading once the M4 was opened and near Exeter at Haselbury Plucknett, Somerset, and my most picturesque addresses, both in Devon, The Fishing Lodge, Deer Park, Weston and The Old Manor, Talaton.

These geographical foci all lie on the western side of south-central England, which goes a little way towards offsetting the fashion in which much of the secondary literature leans towards eastern arable England or the Midlands triangle with its impress of Parliamentary enclosure. Subjects that bear on local studies, or more pertinently draw on them, include agricultural history, agricultural economics, historical geography, landscape history, environmental history, political history and economic history. If this array seems bewildering, it should not do so, because it all relates to the development of rural society and it all inter-connects. It is a means to the ends of investigating historical problems and integrating half-remembered elements from the past.

The book therefore rests on a lattice of local histories from the counties listed, notably Andover and Hampshire more generally; the Oxford district; and Fairford, Gloucestershire, at the junction of the Upper Thames and Cotswolds, where I now live. Wiltshire is added because it is accessible from both Andover and Fairford, and is of special interest to me by virtue of family origins and my association with the Richard Jefferies Society (Jefferies was the first author to coin the term 'wildlife').[3] My doctoral thesis was partly about Herefordshire, and, although it is not convenient to cover that county here, its differences from south-central areas did expand my acquaintance with rural landscapes. Much of the knowledge was aided by bird-watching, an excellent pastime for getting into obscure corners of the landscape.

Additional locations were brought under my notice through repeated visits to naturalist friends who really knew the history of their patch and had, unassisted, earned for themselves the title of environmental historian: Colin Tubbs at Lyndhurst, Lew Lewis in the Newbury district, Patrick Dillon at Great Shefford and Ashbury and Alan Albery at Pamber. These are (some sadly were) people who declined to be pigeonholed by academic discipline. Colin and I often debated what the subject of our interest should be called, whether historical ecology or ecological (now known as environmental) history. We decided that the former meant biologists looking at the development of ecosystems, the latter historians looking at humanity's impact on and exploitation of the landscape and wildlife. Colin was professionally an ecologist, I was an economic historian, but our joint interest was the historicised landscape. We published some papers and he wrote *The New Forest: An Ecological History* in 1968.[4] But Colin's work was ignored even by a prominent historian in the next county, who claimed that in 1970 environmental history had not been invented![5]

Seeking the genius loci of one's haunts, or claiming they are numinous, is no longer fashionable. Whereas my scholarly training certainly leads me to disregard such indefinable terms, I have enjoyed the places mentioned and empathise with people to whom they offer ethereal inspiration, such as the poet Peter Levi's achingly beautiful descriptions of the Oxford district in *The Flutes of Autumn*.

Agriculture's Prominence

The farther back one probes into the economic history of England, the greater the role of agriculture and hence of rural life.[6] The earliest national figures are the social statistics estimated by Gregory King in 1688. They show that agriculture employed 70–80 per cent of the labour force, although the economy was not then fully specialised and many people had secondary occupations. This percentage was already lower than anywhere except the mercantile Netherlands by any reasonable measure must be termed substantial. The proportion of farmworkers continued to shrink but for centuries to come agriculture remained the largest single occupation. Aggregate statistics are few but until the late nineteenth century agricultural land, labour and capital constituted a very large proportion of the economy, even allowing for the fact that the numbers are not always comparable because they sometimes relate to the whole of Great Britain and sometimes include Wales jointly with England. One of the most telling statistics is that into the 1830s land and farm capital amounted to 63 per cent of Great Britain's national capital. More recent attempts at calculation have been hampered for lack of new sources and have not diminished the immense relative size of agriculture before the nineteenth century, when industrial capital began to grow exceptionally fast and farming suffered from the Great (arable) Depression at the end of the century.

The orders of magnitude in the figures are acceptable enough and of self-evident importance for national wealth and well-being. Inequality of ownership was marked: at the 'New Domesday' in the 1870s over half of England was owned in estates of over 1000 acres by as few as 10,911 people. Parts of the estates were given over to the owners' pleasure while the everyday business of agriculture was carried on by tenant farmers. Outside this system lay the remaining privately owned land, in addition to which as late as the mid-twentieth century common land amounted to 1.5 million acres in England and Wales.[7] This was the residue after the centuries' long advances of enclosure and encroachment. Although constituting only 2.5 per cent of the total acreage it was still a large absolute area, widely dispersed through the two countries. Commercial agriculture's economic prominence, coupled with the hangover of communal organisation, make the land sector of outstanding historical importance and a centrepiece of attention in this book.

Angles of Approach

Ground evidence hints at how old methods of exploitation operated, and what natural resources were available for the mere effort of collecting them—hard scrabble tasks involving arcane skills that are difficult or impossible to deduce from abstract principles. John Hughes got this right when he pointed out to sceptical fellow economists that the tools of neoclassical economics do not explain everything.[8] The world is too idiosyncratic for that. The aim of economics, as Hughes knew as well as anyone, is to learn just how much of the variance in behaviour can be accounted for by price theory, which I have heard suggested is about 60 per cent. What, then, of the other 40 per cent, all those tiny disequilibria that economists sweep under the carpet? If this is too formal a portrayal of what really happens at the hands of economic historians, who are more attracted to everyday features than are professional economists, they too play up the main currents and market fluctuations of the past. The force of little things, like those resurrected in this book, is easy for anyone to miss.

A telling parallel or forerunner is found in the work of Joan Thirsk, who had not the least interest in received 'theory' but had worked at Bletchley Park and was no backwoods antiquarian. When she turned to historical research she scoured documents about the agrarian past, following up indications about minor products that she kept on noticing but which others had skipped over. She did not rely on well-known sources or construct series about 'major' changes but put her findings together to portray the underside of agriculture.[9] The most conspicuous movements usually reported of the land sector reflect the preoccupations of late eighteenth and nineteenth centuries farming writers and by extension those who have been over-influenced by them, whether subconsciously or not (*mea culpa*). What other scholars had missed through looking too fixedly at the big, 'capitalist' farmers and the best-recorded statistics turned out to be a sizeable chunk of the rural economy, especially before, say, 1750.

Thirsk discussed less familiar crops and plants, anything that was tried out in the hope it might become profitable. Her work is an antidote to what is in effect the present-mindedness of agricultural historians who skate over, if they ever acknowledge, the intricate labour expended on the infinite productions of the natural world. The secondary literature continues—understandably in the case of textbooks—to be keyed to the main topics that attract the current history profession. There, the attitude seems to be that individually minor products which are hard to quantify must accordingly have been trivial. This is far from correct, certainly in aggregate, as will be shown here with respect to heathland among other topics.

Academic approaches to environmental history are now broadening but biologists have never been very interested in human history and historians are not typically attracted to biology. Despite environmental history's emergence as a sub-discipline in its own right, general histories have yet to integrate it as a matter of course. Consider an edited book on the seventeenth-century county community which ignores economic and agricultural history and reaches the very last page before declaring that studies of topography's relationship to the economy, society and politics might well be revisited along the lines of Hoskins, Everitt and Underwood— and that such a study might be called 'environmental history'![10] The three authors named had all come up with creative theses integrating disparate facts and ideas; the profession has spent the last thirty years trying to amend or demolish their theses, which would be more fitting if equally broad alternatives were put forward.

More remains to be done if believable pictures of past landscapes and occupations are ever to be painted. Thirsk had first-mover advantage; later work has been less comprehensive and less original. Yet when all is said and done, her research was based on documents and printed sources. Investigations would benefit by adding scrutiny of the ground and considering ecology, besides drawing on insights that older inhabitants of the countryside can sometimes provide. In that respect the approach of the

American historical novelist, Conrad Richter, has always seemed a model. Rather than dumping library notes into his books, he deliberately quizzed old people in order to gauge the spirit of the times when they were young. He became aware that ordinary people in the past had not owned much furniture or many household goods; their surroundings were Spartan, like their speech and values. It was the same in England, contrary to the televisual world where everybody is dressing up all the time—as opposed to the reality of wearing cast-offs and with no one fashion ruling at a given time. The much-cited probate inventories seem to offset this bias but may give an unbalanced impression because the median inventory catalogues the possessions of people who were still some way from the poorest of the poor.[11]

How can the lost world of the past be penetrated? The short answer is that it can be retrieved only elliptically. Talking to old countrymen and women is one approach. The oral histories of Suffolk by George Ewart Evans, only a decade older than Joan Thirsk, show what can be done. His compilations of village lore reached into the past by analogy, after the fashion of anthropologists who deduce human behaviour through interrogating remote tribal societies. Admittedly there are limits: the memories of old people and the knowledge they gleaned from their parents and grandparents can take us only so far. Country people may have authentic rustic accents but their experience does not go back inordinately far. The oldest informants I have talked to were a Hampshire farmer and a Wiltshire farmer, both of whom could just recall as children the dreadful harvest of 1879. About that date, Richard Jefferies was tapping memories from the Napoleonic War years, two generations before his own time but he was already too late to make contact with periods before that. It might seem otherwise when reading of the sequence of acquaintanceships certain people claim, hopping and skipping back from individual to individual to someone touched for the King's Evil, but those boasts are symbolic and without much content. Old dialect terms may persist but do not amount to narratives about earlier times. Yet the tune of the past still registers; the pace of life, its laboriousness and its inconveniences emerge to a degree from the reminiscences of people who can 'feel' earlier times because they worked on the land during the threadbare years of the interwar depression. Wood-carvers who replicate the strokes and methods of medieval craftsmen claim even more plausibly to enter the minds of their predecessors.

We must try to capture the tone of the past and round out what less formal sources and physical traces tell us. It is important to steep ourselves in rural ways and make sure to observe such things. I was struck when reviewing a collection on agricultural history which emphasised the storage of grain that none of the authors seemed to have seen or heard of granaries raised up protectively on staddle stones, of which in one county alone 300 remain today. They were looking through the wrong end of the telescope and remained unaware of a formerly routine method of pest control. The practices of less commercial societies offer helpful views. Following the fall of communism in 1989, farming in Eastern Europe gave insights into methods abandoned in the West. In the 1990s, when I lived on a farm at Wantage, Oxfordshire, the farmer could not find an English machine capable of reaping the long-straw grain he grew for thatching. He had to buy a newly made but essentially old-fashioned one from Poland. Even with this machine the only people who would work for him all day on the long-straw harvest were retired farmworkers and other old men. Younger men would not join in. It was testimony to how laborious farm tasks had once been.

Purely documentary and written sources will not take us as deeply into the past as we may wish to go. Landscape evidence and residual practices simply must be included. Having worked on the one-time agricultural significance of gorse (as the chapter on heathland shows), I was chastened to read Chris Howkin's *Gorse, Broom and Heathlands* (2007) in which he teases out the history of cultivating gorse and broom on the Surrey heaths. Faced with patchy evidence in a district where the original commoners' economy had been ruptured by enclosure, undercut when coal replaced traditional sources of fuel and half-obliterated by the mass planting of conifers, Howkins misses no line of enquiry that might clinch what used to take place. He uses signs in the landscape, knowledge of botany, knowledge of chemistry, old testimony, old tools, old documents, old maps, everything, including above all a workaday intelligence that combines and makes sense of the fragments.

Finding something new in a landscape about which so much has been written was hard even in the 1950s and 1960s. By 'new' is not meant something no one else has ever seen but a site or process that had been studied only piecemeal, as if it were a single, unrepeated instance never elevated into a class of phenomena with common roots and broader significance. To illustrate how difficult it is to find something which falls into this category, I claim to have come upon only three examples myself:

the sites of previous use denoted by banked-and-ditched gorse enclosures in the New Forest; the locations of village and town fires which became visible only on noticing that whole blocks of buildings dating from certain early periods were simply missing and the innumerable examples of 'road capture' where private interests had forced awkward diversions and bends on the road system. These finds were unexpected, not least in their wider implications. By discussing traces on the ground and the remnants of dying arrangements I hope to show more broadly here that we have been neglecting vital means of assessing the past. At the risk of giving the impression they are isolated examples and not fungible—not repeated over wide areas—the examples are deliberately drawn from on the spot experience. A little personal material in most chapters will indicate how I alighted on the instances, before debating them at greater length. The autobiographical element seldom stays in the foreground. Its aim is to introduce an element of realism which is not thought appropriate in the stylised formats of academic articles. Showing how I went from the ground to the documents and literature and back again underlines the value of simple observation.

Topics and Themes

Writing this book has impressed on me that overlooked features and obscure corners of the landscape can indeed contribute to broad historical understanding, above all of economic history. The chapters deal with specific types of landscape and resources, typically from the starting point of what was learned from remains on the ground. They lead in a variety of directions, some of them unforeseen in view of their prosaic origins in local field work and ground evidence. The themes include reflecting on how once-prevalent communal systems of land management worked. One chapter concentrates on the unfamiliar, but formerly widespread, variant of such arrangements called lot meads, where rights to mow the hay were reallocated every year in ceremonies involving the drawing of tokens corresponding to strips of grass. By raising questions about institutional change this shows just how far the prompting of landscape features may lead.

The opening four chapters are background. After the introduction in this chapter, the second chapter deals with physical sources installed in the landscape, that is, artefacts. After that, comes a discussion of documentary sources. The fourth chapter advises taking care not to assume, however subliminally, that the countryside of one's youth was the product of ageless, unidirectional trends. A common tendency is certainly to take the context as given, without thinking more about it. But as soon as it is realised that the local scene differs geographically from others, an awakening from unconscious reverie can happen quite quickly. The observer's vantage point is soon located in space. What is less apparent, at least at the conscious level, is a dimension beyond the geographical, which is to say the chronological.

Everybody is aware that their time is different from that of their parents or grandparents. It is taken for granted, without further ado. The landscape may be understood at the back of the mind as a result of historical processes but reflection is needed to fix the phases of change that culminated in the scene before one. Even with a rather specialist education in geography and economic history, it took me some time to recognise the stages that had created the countryside in which I grew up. The realisation was obscured by the brooding hangover of the Second World War and in the Hampshire chalk lands by the peculiar dominance there of prehistoric archaeology over other approaches to landscape history. The bits of the jigsaw were there, they just needed the penny to drop. As Goethe said, 'All the pieces in the hand, lacking, alas! the inner band'.

Before struggling to understand the phases of agricultural prosperity and depression that had succeeded one another through the years, I needed to grasp the significance of the moment when I became an observer. Eventually it dawned on me to ask whether rural history (or its North West Hampshire segment) had experienced a crucial shift. The conclusion was that it had, because in the mid-twentieth century it was fast leaving behind centuries of sheep-fold, horse plough and staffs of forty men on any middle-sized farm, to enter the world of ever-larger farm machinery driven by ever fewer hands. In some ways the question is artificial because the commercial agriculture of lowland England has been changing for centuries. Yet there really does seem to have been a remarkably abrupt transition when long-lasting methods and ways of life gave way to more mechanical, more suburban trajectories. Friends of my own age and interests agree. The transition impressed itself on us because, as naturalists, we were reluctant witnesses to the brutal effects of subsidised, mechanised farming on wildlife and the environment. The fourth chapter, Post-war Time Shift, examines this change in detail as it arrived on the Hampshire chalk.

After these preliminaries, eight chapters follow under the heading of Localities. The first of them, Chapter 5, continues with the chalk land theme. The chalk grassland had been aggressively ploughed in the mid-nineteenth century. During the subsequent depression much had reverted to rough grass, only to succumb to a further assault by the plough during the Second World War. The most attractive old turf was in down-land spots that had been kept as long as possible to provide supplementary feed for sheep folded on ploughland—penned at night within hurdles—for the sake of fertilising cereal crops with their dung. In the 1950s the 'hot spots' on the chalk, as naturalists thought of them, were the ghosts of this old arable sheep system. Chapter 5 records my dawning consciousness that, instead of being timeless, the surviving down-land habitats were 'degenerating' hangovers. They were not ecologically static. They were a transient stage between down-land grass and an eventual covering of scrub woodland. It was a process, not something which could be perceived at any single moment—and only a naturalist's or conservationist's value judgement could think one ecosystem 'better' than another.

Primed by an interest in land reclamation instilled by the chalk lands, I was ready to find in the gorse brakes of the New Forest— that showpiece of Hampshire ecology—suggestive evidence of agricultural advances into the heaths. The incursions were on Crown Land and no one seemed to believe that any reclamation could have happened there. Chapter 6 develops this theme. Heathland in other districts had been more obviously exploited for a whole range of products; they were a distinct sub-system of the rural economy which is also discussed here. But in the early 1960s the heaths were not easy to reach from Oxford, where I then was; fortunately another jewel of environmental history conveniently presented itself in the Thames valley. The essence was what proved to be the expiring years of a medieval ceremony at Yarnton, where the rights to mow strips of hay were allocated by the drawing of lots (Chapter 7). Botanical variety here and in some neighbouring meadows was high but the apogee of interest lay in the ceremony itself, one of the last exponents of a method for allocating resources that had been standard in the river meadows. Two or three comparable arrangements were to be found, including one in Devon which revealingly was not all it seemed, yet communal institutions proved even more fragile than the natural history which in the 1960s drew the little attention then granted to these areas.

Oxford led me to another habitat in the district which was better known than lot meads because it was a popular resource for walkers. This was the Ridgeway, a drove road running along the northern escarpment of the Berkshire Downs. Its historical interest lay in funnelling sheep flocks to the great market halfway along at East Ilsley (Chapter 8). Already drawn to the subject because the even larger sheep fair at Weyhill was only a couple of miles from Andover, I began to think about the role of droving in the agricultural economy. Long afterwards, Patrick Dillon and I wrote a book about Ridgeway country, which we conceived as a long, high, narrow and distinctive ecological 'island'.[12] Part of our interest was sparked by the arable take-over of hill country like the Berkshire Downs along which the Ridgeway ran (Chapter 9). Environments of that type have undergone striking land-use changes over the centuries, almost oscillations, and evidence of previous phases of occupation remains visible despite the near desert created by modern industrial agriculture now operating from major farmsteads some distance away.

At this point attention switches to Gloucestershire, where I now live, with Chapter 10 covering the park at Fairford in the Cotswolds. Parks were a conspicuous feature of the English landscape and it was tempting to report on Ashdown Park on the Lambourn Downs section of the Berkshire Downs. That was a great sporting estate where the Cravens, who owned it from the seventeenth century, preserved more rough ground than might have been expected only eighty miles from London. They could afford to let farming take second place. But Ashdown's history is part of the hill country story and figures in Jones and Dillon, *Middle Ridgeway*. The focus here is instead on Fairford Park, a pleasure park with a basic eighteenth-century layout. With a little effort much of the apparatus of the landscaping and estate management of that period is discernible and the owners' one-time sway over the little town at their gates continues to be apparent.

The final two substantive chapters discuss the history of resources. One of them, Chapter 11, shifts from treating a single location to tracking the national expansion of fodder supplies. It considers how grass feed was supplemented by additional crops and is aimed at reinforcing Chapter 9, because the developments were of special moment in colonising the bare hill country. The other resources chapter (Chapter 12) is on Gloucestershire, Cotswold-based, and deals with two very different commodities, wool and wood. It reveals the long history of how cloth making was replaced over much of the county by wool-stapling which, when it too collapsed, was unexpectedly replaced by wood-working. The aim was to supply Birmingham gunmakers with gun-stocks of beech to meet the heavy demand for cheap weapons in the African slave trade. The Cotswold beech woods also came to support other types of wood-working and even introduced steam engines very early. This is an atypical case where ground evidence is almost non-existent and needed to be exhumed from the most minor documentary and published references.

The bulk of the localities studied throughout the book required combinations of field work and library work to discover how each was occupied and exploited. Much of that history, although by no means all, relates to the agricultural sector, as well as indicating how remnants of past ecosystems had come to be formed in the first place. A chance observation on the ground may set the chase in motion but sometimes a published reference sends one out to look at the ground anew. The combined approaches help to paint a fuller picture of the course of economic development.

Notes

1. Other types of habitat have been dealt with elsewhere: coastal areas and rivers in Eric L. Jones, *Revealed Biodiversity: An Economic History of the Human Impact* (Singapore: World Scientific, 2014), chapter 7, 'Landscapes of Destruction: The Sacrifice to Trout' and chapter 9, 'England: Reclamation and Exploitation', and wetlands also in Eric L. Jones, *Barriers to Growth* (Basingstoke: Palgrave Macmillan, 2020), chapter 13, Floods.
2. Kibworth was the site of a truly outstanding television series by Michael Wood first broadcast in 2010.
3. *Oxford English Dictionary* citing Richard Jefferies, *Wildlife in a Southern County*, 1879.
4. Colin R. Tubbs, *The New Forest: An Ecological History* (Newton Abbot: David & Charles, 1968).
5. Christopher Taylor, 'Dorset and Beyond', in K. Barker and T. Darvill (eds.), *Making English Landscapes* (Oxford: Oxbow, 1997), p. 20.
6. See Phyllis Deane and W. A. Cole, *British Economic Growth 1688–1959* (Cambridge: Cambridge University Press, 1962); B. R. Mitchell and Phyllis Deane, *Abstract of British Historical Statistics* (Cambridge: Cambridge University Press, 1962); and Dominic Hobson, *The National Wealth* (London: HarperCollins, 1999). The present book is mainly concerned with the seventeenth, eighteenth and nineteenth centuries.
7. W. G. Hoskins and L. Dudley Stamp, *The Common Lands of England and Wales* (London: Collins New Naturalist, 1963), p. xv.

8. J. R. T. Hughes, 'Fact and Theory in Economic History,' *Explorations in Entrepreneurial History* Second series 3/2 (1966), pp. 75–100.
9. Joan Thirsk, *Alternative Agriculture* (Oxford: Oxford University Press, 1997).
10. Jacqueline Eales and Andrew Hopper (eds.), *The County Community in Seventeenth-century England and Wales* (Hatfield: University of Hertfordshire Press, 2012), p. 136.
11. Cottagers tended to have more cheap ornaments than they had furniture.
12. Eric Jones and Patrick Dillon, *Middle Ridgeway and its Environment* (Salisbury: Wessex Books, 2016).

CHAPTER 2

Sources: Artefacts

The primary evidence for environmental history is the landscape itself, as subsequent chapters will show. It needs however to be accompanied by the findings of ecology or natural history, which has a long tradition in the southern counties (think Gilbert White in Hampshire). Early works on wildlife were perhaps too prone to concentrate on numerical distribution, which meant cataloguing shot specimens and only incidentally revealing ecological connections. Happily that has changed. Topographical histories also sprang from a rich seed-bed (think John Aubrey in Wiltshire) but earlier ones also had a tendency to let themselves be tugged off-track, in their case into the sidings of antiquarianism. Nineteenth-century histories tended to dwell on medieval topics such as manorial descents and ecclesiology, sometimes adding (when local circumstances made it hard to avoid) a glance of the exploitation of resources, such as brine salt on the coast. But only *en passant*.

Take as representative Hampshire, for which an early comprehensive history was T. W. Shore's *History of Hampshire* of 1892.[1] Shore concentrated on early periods, especially the Middle Ages, and his book gave little indication of the breadth of his lesser productions. His pieces on Hampshire parishes in local newspapers nodded towards geology, land tenure, forestry, boundaries, place names and so forth, but he must have judged what an educated readership expected in his day: a diet of medieval

E. L. Jones, *Landscape History and Rural Society in Southern England*, Palgrave Studies in Economic History,
https://doi.org/10.1007/978-3-030-68616-1_2

history, stodgy to modern eyes, only lightly seasoned by his personal interest in science. He placed more serious articles on an impressive range of subjects in the *Papers & Proceedings* of the Hampshire Field Club, but Shore was a geologist rather than a naturalist and did not press the relationship between habitat change and living organisms. The days of the great generalists like Aubrey, or Robert Plot in Oxfordshire, or pioneer agricultural writers like Edward Lisle in Hampshire and Wiltshire, were not readily recaptured and only after the Second World War were (still sporadic) efforts really made at integrating the strands of learning.

Environmental history was thus a long time swimming against the tide. Given the range of possible sources, that is a pity. Besides features of the historicised landscape, it can draw much from physical artefacts, since England is an antique shop, or junk shop—or a treasure house. Despite decay and positive orgies of destruction, the country remains crammed with objects from the past. There are said to be one million Georgian houses and over 20 million gravestones.[2] Enticing though this may be to the historian and archaeologist, the profusion has its downside by obscuring the fact that economic development must continually accommodate ancient landscape patterns, constrained in turn by equally antiquated property rights, and is obliged to cope with the massive changes brought about by modern building. The geographer's term, palimpsest, describes the succession of landscapes inscribed one on the other; no label could be more appropriate to so long-settled a country. When he described the making of the industrial landscape, the historian Barrie Trinder realised that mines, canals, railways, factories and towns were already obliged to accommodate themselves to far earlier distributions. He listed these as boundaries at least as old as the Anglo-Saxons, medieval monastic estates, burgage plots laid out in the planned towns of the twelfth century, and common land unenclosed, he thought, because the lordships of manors had been fragmented by the Wars of the Roses.[3]

Asked to explain why English economic productivity is so low, a Treasury study could point with confidence only to the way in which distribution costs are raised by a network of communications originating infinitely far back. The road system is inadequate for modern traffic and, in the face of so many obstructions, legal as much as physical, replacement would be insurmountably expensive. Developers, crass though they may be, have no choice except to work around the myriad obstacles. Much of England's infrastructure and built environment can be heaven for the antiquarian but hell for the entrepreneur. The question is, is the glass half-full or half-empty?

In environmental history the evidence of the ground is primary. Areas of low intensity land use are likely to be the less-disturbed districts or corners of the countryside, such as undrained marshes or stretches of the coastline that provided a living of sorts for net fishermen. Again, it may be derelict coppice woodland that was once integral to the agricultural economy but lost its *raison d'etre* when hurdled sheep ceased to be profitable. Other sizeable patches of land now under less than maximum use include old ornamental parks whose owners prefer amenity to full potential income. Traces of former practices may not be obvious because they are obscured by later ones or may be present because such areas have been virtually deserted as no longer profitable. National parks, for instance, tend to be areas somewhat out of the economic mainstream, while nature reserves may display methods of exploitation well behind current best practice or which were aimed at now outmoded natural products. Alternatively they may present the scars of old types of exploitation, including quarries, spoil heaps, boundary banks or former roadbeds.

The point is sometimes made that entirely marginal habitats yield fewer relics and 'cultural hotspots' than once-prosperous districts that have since stagnated.[4] This is approximately consistent with the presence of former ploughing ridge-and-furrow 'fossilised' today under grass paddocks close into settlements, when further into the fields the remains of cultivation have been ploughed right out.[5] It is also consistent with the observation that antique farm implements tend to survive better near towns, where change came early and gradually, than in completely rural areas which were altered late and rapidly. The sites described in this book were, however, mostly quite distant from settlement, at least in English terms—land that had been occupied by very prosperous agriculture only from time to time. The modern abandonment of such areas is what has preserved the remains of their former cultivation and through lack of disturbance has simultaneously kept them interesting to the naturalist and conservationist.

The land may be cherished by its modern occupants for its amenity value, while in so densely settled a country developers are likely to be eager to change its use. Conservation has an opportunity cost. Many of the most attractive rural landscapes support a tourist trade and nostalgia industry (commonly obsessed with the Tudors), whose practitioners resent restrictions on commercial development. What has hitherto succeeded in protecting versions (or simulations) of past landscapes are planning regulations designed to limit changes of use, plus a somewhat rigid 'listing' of historical buildings that are meant to remain unaltered. As a result land use tends to evolve in a piecemeal fashion, with production sometimes forced into second-best places.

The choice of protected areas is not always imaginative. It derives from conventional appreciations and tends best to defend the bigger battalions among historical monuments. It must be conceded that preserving everything and everywhere that might prove attractive would be impossible on grounds of practicality and cost. Needless to say planning is a legal minefield, with those wishing to amend current uses perpetually chafing at restrictions they see as fossilising the countryside. Developers by no means lose the case for change every time. Ultimate decisions rest in the hands of inspectors who do not necessarily possess a broad educational background or sympathy with preserving the historicised landscape. A recent example concerned a proposed housing estate at Coate, Swindon, where the inspector was unfamiliar with the works of the leading Victorian rural writer, Richard Jefferies, whose home had been Coate farmhouse. Ignorance or incuriosity of this order is also common among local and county councillors. This is not to urge that the claims made by preservationists are always realistic or accompanied by suggestions for the alternative development sites that are needed if people are to be housed as well as monuments preserved. History and modernity do not sit comfortably together and the planning system seldom satisfies every interested party.

The present chapter and the next are introductory and consider forms of evidence supplementary to the landscape in two categories: here, physical structures or artefacts, and in the following chapter, paper records. Start with artefacts. Among physical structures the most celebrated are ecclesiastical buildings. Churches are almost but not quite ubiquitous, their number finally decreasing now that Anglican congregations are shrinking. Admittedly the starting point was high: G. G. Coulson is said to have reckoned that medieval England had one church per one hundred families. The interiors, from which so much history may nevertheless continue to be deduced, suffered badly during the Reformation, to a lesser extent during the Civil War and Commonwealth periods, and as a consequence of ill-conceived 'restoration' at the hands of the Victorians. Many suburban churches were pulled down to give fields of fire during the wars but at least they could be rebuilt. How much more of the stock of religious buildings might have vanished had the Restoration never occurred is a moot point. Even the 'massive cavern of stone' which is Gloucester Cathedral had come close to demolition in 1657. The demolishers had already agreed who was to have what proportion of the plunder and had their implements and tackle ready to hand when they were stopped by the borough authorities. Gloucester was saved and taken together cathedrals and parish churches, neglected as many are today, still hold an unimaginable store of historical memorabilia.

The next most visited and admired buildings are country houses. They were the centres of estates on which the organisation of rural England was based. It is astonishing how many such houses survive given the scale of the losses that occurred, especially in the fifteen years following the Second World War, when many had been contemptuously treated as military billets. The losses are eloquently deplored by the architectural historian, John Harris, in *No Voice from the Hall* (1998). His hobby as a youth was to break into empty mansions. There he found costly furnishings and exquisite cabinets of curiosities in disarray and disrepair, with estate papers scattered about, including in one case the working papers of a famous mathematician. Sometimes houses which their owners had abandoned or failed to reoccupy had descended into the hands of philistine farmers, who used the most ornate rooms for storing sacks of produce or housing chicken and sheep. At Whittington, Gloucestershire, one of the prominent Gloucestershire Arkell family actually sawed off the newel post of the stairway to make hauling up sacks of grain easier. It was not the only example.

Another architectural historian, Derek Sherborn, likewise lamented in *An Inspector Recalls* (2003), the long saga of post-war country house demolitions. He gives scarifying details of the philistinism and incompetence complicit in the wave of destruction. Correct though he is about official vices and chicanery, the catch is that his criterion of significance is entirely his own aesthetic opinion. Other architectural historians may agree with him but that is not evidence. The historical circumstances of the depressed interwar years, the war itself, and the bleak era of rationing that followed should all be considered. The likely cost of repairing, restoring and preserving every fascinating building finds scarcely a mention and two pressing problems of the 1940s and 1950s, the financial embarrassment of owners and the chronic scarcity of materials with which restoration might have been carried out, receive from Sherborn the most fleeting acknowledgement. Rehabilitating this one segment of the country's heritage would have necessitated a massive diversion of scarce public resources. However sympathetic one may feel about the destruction of historic buildings that would have been unjustifiable. More remain than any one person can visit.

Other waves of destruction have occurred more recently, self-inflicted wounds one might say. Councils have shut down libraries and sold their rare books, which is as if the use made of them at the present day is an adequate measure of their potential in perpetuity. Over what period should a return be expected—an intellectual return—from historical sources of this type? One might as well sell off the Bodleian Library on the grounds that at any one moment the dons and students do not have their heads down making sure the bulk of the stock is being read. Councils have also closed museums and dispersed the contents, as if flints and fossils are pointless junk and stuffed birds are distasteful in themselves. Destroying stuffed specimens and egg collections was always unimaginative and proves misconceived now that techniques have been found to extract from eggshells and feathers information about scientific matters such as climate change and pollution. The classic example was the late Derek Ratcliffe's discovery that the shells of peregrine falcons' eggs were thinning, and the species declining through agricultural applications of DDT; he could not have demonstrated this without reference collections of birds' eggs.[6]

Physical objects are dumb records and to interpret them, and the landscape itself, additional evidence is commonly needed. Interpretation benefits from manuscripts, paintings, maps, photographs and aerial photographs, or snippets of detail tucked away in old books and articles. Among all these sources, paintings are the most ambiguous, since attributions are sometimes doubtful, there are many fakes and forgeries, and artists are capable of rearranging scenery for the sake of pictorial effect. Landscape paintings are likely to be most reliable when different artists have rendered the same scene or when the making of a faithful record was their expressed aim, as it was in the 'Recording the changing face of Britain' venture of the early 1940s (1500 illustrations in the Victoria & Albert), which was a less well-known equivalent of the work of the war artists.[7]

Despite the erosion of artefacts and buildings by moth and rust, and sometimes by needless destruction, the sheer number of structures now with some measure of preservation is breath-taking: churches, cathedrals, castles and listed houses. They are not invariably the categorical representations of the past they seem, because in the long-run architecture is surprisingly fragile, liable to be mutilated, 'improved' or simply neglected. I have suggested elsewhere that the layout or plan of a settlement promises still greater antiquity than the houses standing on it.[8] For all that, buildings remain records of incalculable value. England is a rich country where Georgian houses alone are more numerous than those of a similar period in any other country.

Agricultural buildings are less often analysed by architectural historians, whose interests tend to lie more in the arts than social history, but farms and barns do attract some specialist attention. The interest is usually in the structures, especially of buildings in farmyards (despite restoration projects in some counties), although this professional concern seems insufficient to ensure the protection of even the most interesting buildings.[9] There is less attention to the geographical distribution of farmhouses. The pattern of farmsteads can be revealing as a means of tracing the spread of permanent cultivation onto downs, wolds, heaths and former commons, movements which radically altered the landscape. In addition, old farm equipment fills the museums and old tools fill the antique shops.[10] The attics of private houses continue to disgorge memorabilia of wholly miscellaneous sorts for television's Antiques Road Show and a host of dealers. Enforced leisure for so many people during the pandemic of summer, 2020, resulted in utterly unprecedented numbers of finds, ranging from hoards of gold coins to hitherto unknown crop marks. There are, to be sure, curious gaps in the public collections of artefacts, broadly conceived, stage waggons being an astonishing omission given the popular zeal for restoring obsolete means of transport. Yet the total volume of historical resources is staggering.

Objects are better at raising questions than supplying answers. Physical remains do not necessarily proclaim their original purpose and it may be hard to reach even the first stage of enquiry, which is to assign a date to them. Jack Hargreaves warned that no one really knows the intended use of every disused farm tool found in the agricultural museums that sprang up in the final decades of the twentieth century.[11] Tools were made by local blacksmiths for local customers with local needs that may no longer be understood. Over-confident labelling is a danger. In the face of these uncertainties, besides insensitive land-use planning, excessive ploughing and great losses of material, it seems almost Panglossian to point out just how much usable evidence does survive. Nevertheless, the jigsaw has plenty of missing pieces and to fill in as many gaps as possible we must turn to the documents.

Notes

1. T. W. Shore, *History of Hampshire* (Wakefield: E. P. Publishing, 1976 [1892]).
2. Estimates of the houses are discussed in John Woodforde, *Georgian Houses for All* (London: Routledge & Kegan Paul, 1985), p. xiv; the figure for gravestones is from Philip Rahtz, *Living Archaeology* (Stroud: Tempus, 2001), p. 172.
3. Cited in Eric L. Jones, *Barriers to Growth* (Basingstoke: Palgrave Macmillan, 2020), p. 68.
4. H. Renes, 'Historic Landscapes without History? A Reconsideration of the Concept of Traditional Landscapes', *Rural Landscapes* 2/1 (2015), no pagination.
5. A piquant example is at Langford, Oxfordshire, where broad rig waves beneath the village cricket pitch!
6. The use of museum data in ecological research continues. See e.g. Mattias Hagman et al., 'Grass Snakes (*Natrix natrix*) in Sweden Decline Together with their Anthropogenic Nesting-Environments', *Herpetological Journal* 22 (2012), pp. 199–202.
7. David Mellor et al., *Recording Britain: A Pictorial Domesday of Pre-war Britain* (Newton Abbot: David & Charles, 1990).
8. Jones, *Barriers to Growth*, pp. 68–69.
9. Note for instance the neglect of the superb Chazey Court Barn in Caversham near Reading discussed in *Private Eye*, 5 June 2020.
10. Nigel Harvey, *The Industrial Archaeology of Farming in England and Wales* (London: Batsford, 1980).
11. Jack Hargreaves, *Out of Town* (Wimborne: The Dovecote Press, 1987), p. 130.

CHAPTER 3

Sources: Documents

Gaps will remain, because so much was never set down on paper. Mansions and properties can be found that have been in the hands of one family for so long that they have no deeds or conveyances at all but these are very, very rare. Merely finding written sources would not however end the search for former use and occupation of the land. Where formal documents do exist, they can be infuriatingly enigmatic or plain misleading. Manorial court records, says Gillian Tindall, were minutes for people who knew what they meant, 'and had the unlettered habit of keeping all information in their heads anyhow'.[1] Cecil Torr, a lawyer himself, noted that lawyers will copy clauses into documents without considering whether they apply, just as laws may stay on the statute books when they are no longer enforced.[2] He goes on to give examples of copying errors in publications he had studied. Despite every type of accident and omission, material is nevertheless sometimes as voluminous as can be handled. We must do the best we can with what is available, making sure to consult as diverse a set of sources as possible. That includes the work of the old antiquaries whose faults, as Torr found, do not prevent them dropping all sorts of hints and clues.

The formal quality of much documentation was high. The care lavished by craftsmen on material objects was matched by the clerks who kept the

E. L. Jones, *Landscape History and Rural Society in Southern England*, Palgrave Studies in Economic History,
https://doi.org/10.1007/978-3-030-68616-1_3

records of businesses and estates, and whose penmanship was mechanical in its consistency. Their effort was repetitious and could even be routinised: Richard Jefferies says that already in the 1860s, with so many legal items needing to be transcribed, the leading solicitor in Cirencester farmed out work to be copied in London.[3] John Moore, who fled from his family's auctioneering firm in Tewkesbury (established in the 1750s), says that as late as the 1920s it was using morocco-bound books rather than 'modern' tear-off sheets, because of the belief that, 'if a man bought Lot 224 Bedroom Utensil and sundries for two shillings the fact should be recorded for two hundred years'.[4] Ledger after ledger was on the shelves in a back room of the office where I took notes on them in the 1960s.

What can England have been like in 1939? To judge from surviving examples, solicitors' offices and private houses were so stuffed with the paperwork of the ages, it is hard to imagine how people found room to move. It is sobering to think how much documentary material there was, let alone to contemplate the fading eyesight of the hapless clerks who had sat at high stools penning away in Victorian offices. In the library of the Agricultural Economics Research Institute at Oxford there were after the war rich pickings for the historian—but only pickings. The University Printer, that is to say the head of Oxford University Press, was a civilised man and a friend of the Institute's director, Charles Orwin. He used to fish out of the sacks of waste paper sent to the Press in the wartime salvage campaign any especially fine-looking documents and pass them along. The documents were all out of context. They were fun to look at even though they were merely jewels ripped from the settings of family or farm or estate archives, the remainder of which had been pulped in a massive recycling.

The Cirencester and Tewkesbury records are both now in the Gloucestershire record office, with innumerable office files accumulated by long-gone estate clerks. Other hoards have not been so lucky. Fire is a perpetual hazard, as is deliberate or unthinking destruction. James Lees-Milnes' biographer remarks circumspectly that the National Trust 'has not been remarkable for conserving its own archives', dispersing them among various repositories where many have gone astray.[5] If a body dedicated to historical preservation could behave so cavalierly, lapses elsewhere can readily be believed. The catalogue of the Berkshire Record Office observes of some deeds that they 'were saved from documents thrown away by a firm of solicitors' and presented to the archives by a Marlborough antiquarian. The deeds referred to the period 1590–1798. How educated people could have dumped them defies belief, especially in view of the common practice among solicitors of hanging on to material against the chance that someday it might come in handy. They are likely to be right about the long useful life of particular documents; there are firms of solicitors whose first resort in boundary disputes is the enclosure award. From time to time one finds county councils referring to the awards with respect to rights of way. I can think of a recent example where documents concerning an enclosure of 1738 were consulted for the purpose.

In other quarters ignorance has often prevailed. A builder whom I knew in Somerset rescued a pile of papers from a bonfire to which his own brother had consigned them. The well-made chests storing the documents were being emptied out for their employees to keep tools in. The documents began with the Middle Ages. On a Gloucestershire estate the agent told his staff to burn all non-current paperwork, as I learned from one of the men who had kept back as a keepsake a volume concerning the construction of the local railway. Something similar happened on a Hampshire estate where the mansion was being demolished. One of the workmen removed a ledger as a souvenir. I told the county archivist discreetly about this, the landowner was contacted, but swore all the documents had been deposited: they had not.

Despite the expunging of so much of the past by accidents and assaults, a surprising number of manuscript collections, pearls in a historian's eyes, succeeded in escaping. In 1958 I was recruited to write a thesis about the Hampton Court estate, Leominster, Herefordshire, bought in 1809 by Richard Arkwright, son of the cotton-spinner Sir Richard Arkwright. Studies of individual estates were fashionable at the time, although Lord McGregor was shortly to point out in an introduction to the sixth edition of Ernle's *English Farming Past and Present* (1961) that it came close to an exercise in futility. The analysis of a collection was likely to take a full-time researcher a couple of years and, with potentially 2000 great estates to investigate, the scholarly task would never be finished. Generalising from the results of the handful which were studied was unsafe because there was no means of telling how representative they might be.

What does one wish to find out by studying estates, the variety of which was considerable? By the early twentieth century, estate owners constituted a class in decline, 'much of its membership becoming decadent or eccentric'.[6] Policies of estate management could be correspondingly wayward. The neatness of columns of figures does not guarantee that they make sense. Modern archivists seem taken up with the administrative problems of managing estate collections and what emerges has a fair chance of being stories about the louche elite rather than analyses of the commerce of yesteryear.[7] Financial records and associated rentals are bulky, comparatively little used, and indeed hard to use, as I discovered when faced with the Arkwrights' ledgers.

Scholars are temperamentally inclined to bury themselves in the documents at hand and see what might emerge from the lucky dip. As a beginner I was ready to enjoy just that. Sixteen tea chests crammed with ledgers were brought to Oxford from the Herefordshire Record Office, mainly because the Office—which was in the county library—was desperately short of space. Overwhelmed by this material, so clearly tabulated and yet so lacking in summary, I looked for a key to its meaning, whatever that might be. I therefore wrote to such of Richard Arkwright's descendants as were traceable, which was not difficult because he had acquired estates for five sons, some of whom had left heirs.

With one I hit pay-dirt. David Arkwright lived at Kinsham Court, Presteigne, an estate to which the family had been reduced since their fortunes took a knock during a downturn in both agriculture and (in 1904) the cotton trade. Previously they had lived in Hampton Court, Leominster, the estate that David's ancestor had bought. To most people Kinsham would not have seemed much of a come down. It had four miles of own fishing, connections with Byron and Florence Nightingale, a collection of pictures of which several were by David Cox, various Arkwright memorabilia, and importantly from my point of view the missing half of the manuscript collection. David Arkwright was most hospitable and invited me to stay in order to work on the letters and diaries that alone made the formal ledgers explicable. I was later able to persuade him to deposit them in a rejuvenated Herefordshire record office run by the redoubtable Meryl Jancey. No one had hitherto approached David about his collection, a lack of interest that is less likely today. In the last few years I have been permitted to browse in a couple of significant estate collections still in private hands, but the archivists do know of them. The problem from their point of view and of researchers is that important owners can be unwilling to make deposits. The rationale of one owner, a solicitor by profession, is that if all documents are gathered into a single repository and there is a fire, a county's entire history might go up in smoke. He has a point because at least one archive office has suffered such a fate.

The private owner is under no obligation to permit access, which may be seriously inconvenient in his or her home. The researcher is also likely to find it awkward to arrange visits to collections in scattered houses. Making journeys to record offices, which are now often bureaucratic appendages of county councils and not necessarily organised to help people from a distance work on their holdings, is difficult enough. Travelling around even the established archives is expensive, time-consuming work, and absent generous grants, tends to bind research workers to one office. This helps to explain the excessive grip of county studies on rural research, where an academically more realistic—although more costly—approach might be to work on the history of a family with estates in more than one county. Historically, the family was the ultimate managerial link.

Reports of mouth-watering prices received at auction for 'old things' persuade some owners that what they have, even the dullest accounts, possess greater commercial value than is the case. They are more likely to be correct when the items are decorative. Amazing sums are certainly realised for bric-a-brac if it has a connection with a well-known person or event. I am reminded of a Danish archaeologist, who when told that a folk museum was opening, snapped, 'rubbish is rubbish even if it's old rubbish'. On the antique market serious research purposes are no consideration. Pubs used to buy indentures unconnected with the house or even the district in order to frame them and hang them on the wall. It was possible to sell parchment to furniture dealers who would cut it up to make lampshades. In the 1950s, these lampshades were commonly found in the more flamboyant pubs, such as roadhouses trying to create their customarily incongruous 'Tudor' atmosphere. And deeds and indentures are still sold as curiosities by dealers, the catalogues and websites of some of whom offer (at high prices) materials that would be irreplaceable pickings for the historians of some places and families. Private sale means they are dispersed and lost to scholarship. Individually a deed may provide only a nugget of information but the ability to scan documents digitally means that large collections might now be analysed; it is unfortunate that a random dispersal of documents is still in train.

At times, too, old books are kept as souvenirs when the rest of a library is dispersed. A friend showed me a signed volume from 1652 with indications by her ancestor of support for Cromwell and Hartlib. It would have been impossible to guess she owned such a thing. Nor should marginal notes in books be dismissed as too scanty or enigmatic to be enlightening; they may seem episodic but when collated have been put to good use in tracking changes in crop species by Mauro Ambrosoli.[8] He was not so much discovering a new source as virtually inventing one where none was thought to exist. That was a real achievement. But Amrosoli's extensive researches involved visiting international repositories to trace marginalia in books written in several languages. Smaller samples may admittedly reveal less of a pattern.

The continued existence of so much material of so many types, relative to what an individual researcher could possibly handle, sits paradoxically with the incremental leakage of material and the great epidemics of loss through wartime bombing and the salvage campaign. 'The greatest disaster in the world of English archives in recent times', proclaimed the economic historian, W. G. Hoskins, 'was the destruction of the probate registry in Exeter in World War II. In one night in May 1942 the entire contents of the registry in Bedford Circus perished forever: probably a hundred thousand wills and perhaps half that number of inventories'.[9] Only a small fraction of this material had ever been copied. Disaster it was, and a historian of Devon like Hoskins was bound to feel it acutely, though it is a moot point whether in the country at large the toll of damage by spring-cleaning and miscellaneous misfortunes has not in the end added up to just as much harm.

On the one hand a range of items vanishes into the antiques market, on the other hand surviving documents are not always cared for and institutions that should know better can fail to look after what they have. The boast of the established church used to be that it placed an educated man (sic) in every parish yet it is still easy to find Bibles with handwritten annotations and alterations to dates literally mouldering in church chests. Early water colours of churches, which may portray the landscape, have all too often been removed for storage or been stolen. Some estate offices are thoroughly offhand about old documents, perhaps because estate stewards belong to what have been called 'the self-engrossed middle classes' and tend to be more interested in property rights, shooting and fishing than in things bookish. I have known poignant examples of loss on estates in Gloucestershire, Hampshire and Somerset. This is another instance where keeping one's ear to the ground can be informative. As mentioned, workmen told to dispose of estate records sometimes filch interesting items and may proffer tips that destruction is on the way.

What remains in private hands is not always accessible and owners may wish to keep quiet about their holdings for a variety of reasons, although others are extremely generous. Gifts can arrive out of the blue.

In 1884 the vicar of Fairford received a book of seventeenth-century documents relating to property in the town, the *Fairford Booke*. The donor, in distant Glasgow, had found it when clearing out the papers of his uncle, a surveyor and conveyancer, and sent it on the off-chance, it might be interesting. Pure luck. Even today, despite all the hazards, finds may be made and I have come across two other unexpected collections of deeds in Fairford while writing this chapter. Discoveries of this sort depend on personal knowledge, being able to present bona fides (local connections being almost indispensable) and maintaining good relations in the community. The same applies when houseowners themselves were unaware of what they had but have chanced on documents squirrelled out of sight by previous occupants. John Woodforde remarked on such finds in the 1970s, adding that, 'householders are recommended to shine a torch when boards are lifted for re-wiring work'.[10]

Institutions connected with well-known personalities or with local ties may, by virtue of having a fixed presence, be in a position to attract gifts or at least offers of material. It might be thought that nothing new was ever likely to surface with respect to the writer, Richard Jefferies, in whom generations of aficionados have shown an interest. Yet during the many years I was on the committee, the Richard Jefferies Society was able to acquire collections relating to one of Jefferies' ancestral families and to the writer (Samuel Looker) who did most to keep Jefferies' name alive during the years when natural history writing was less fashionable than it has become. Reels of a seemingly lost film made in 1950 about Jefferies' haunts turned up in an attic halfway across the country.

My acquaintance with account books did not start with the Arkwrights' estate ledgers. It had begun with farm accounts in the Hampshire Record Office when I was working on my undergraduate dissertation. Account books which were held contained data about crops and livestock that was available nowhere else. The year was 1957. An objection to bothering with farm accounts, which was sometimes forcefully voiced by agricultural historians at that date, was that farmers did not keep records! It is easy to guess that hard-pressed small men did not do so, which may unbalance the available sample, but as a general proposition it is absurd. At that date few, very few, examples of farm records had ever been published. Notable were the seventeenth-century accounts of Robert Loder at Harwell, then in Berkshire. On the head of that one pin an inverted pyramid of interpretation was made to balance, as textbooks of the day testify. Matters are slowly improving and recently some serious analysis of farm accounts has been attempted.[11]

In the 1950s and 1960s certain shibboleths dominated agricultural history. We can leave aside the school textbook insistence that agricultural change was brought about suddenly in the eighteenth century by four 'patron saints' of agriculture, although as late as 1979 my daughter was taught this mantra in a Somerset school. Three other beliefs were also almost caricatures. One was that enclosure must mean Parliamentary enclosure, another that turnips were not introduced until the late eighteenth century, and yet another that the open-field system was flatly incapable of innovations in cropping. About enclosure and turnips the handful of economic historians who specialised in agriculture were better informed but the general run of historians clung to the old tale of late onset change. The relative strengths of change at various periods and in different regions are of course separate and still contentious issues. The point is that farm accounts—farm record books—made demonstrating early change much easier.

The farm record books in the Hampshire archives brought me face to face with a gritty reality. Literally: good Hampshire soil poured onto the Record Office desk from the gutter of one book. Mostly, however, archivists had little interest in farm records and it was a rare office that catalogued them as a separate category. Opportunities to acquire records, once they had been neglected might never come again. The Victoria and Albert Museum has had a 'rapid response collecting' strategy since 2014 but is, I imagine, better resourced than county record offices. In the 1960s three offices actually declined to take documents when they were available (I offered one collection on behalf of my landlady), and in another instance the post-1880 parts of a collection were left on an office floor as 'modern commercial records'. That dismissal was a red rag to the bull to an economic historian.

The paucity of farm accounts in the archives, together with the lack of concern about collecting them and denials that they might be historically informative, spurred me to try to amend the situation. Nuffield College gave me a grant to employ a research assistant whose task would be to visit record offices, identify what they held and make notes on the contents. A secondary aim was to secure accounts held in private hands. The assistant was E. J. T. Collins, who afterwards moved to what was then the Museum of English Rural Life at Reading University. We started out with a splendidly encouraging evening during which we telephoned eight people in parishes around Oxford whose family names had been given as 'farmer' in a nineteenth-century trade directory and received five positive responses. Admittedly in one case, when we did visit, we were told 'you should have come last year because we burned all those old books except this one which we kept as a memento!' That little tragedy was not unique but by and large the finds were promising. We continued collecting at Reading in our spare time.

Summing up the results of a year's visits to record offices (a slow business) and the contents of such ledgers and notebooks as we were able to acquire led to the conclusions summarised below.[12] The results fell into four main categories. One pertained to the home farms of landed proprietors who may sometimes have pioneered technical advances in agriculture but were hardly typical of the industry. A second group comprised accounts of farms taken in hand by an estate—accounts that were kept by farm bailiffs prior to re-letting. These were usually well set out but tended to cover short spells of time. Third were the accounts of large occupiers and the lesser gentry, a progressive group on the whole and once more not wholly characteristic. Given the presumed bias towards the more advanced agriculturists in the previous three classes, the fourth was the most interesting: ordinary working farmers. Of special note is the fact that by the mid-nineteenth century standardised printed ledgers were on sale, which rather gives the lie to the insistence that farmers never kept records.

Viewed as business accounts, farm record books have advantages and disadvantages. The representativeness of any one set of accounts is hard to demonstrate. There are few entirely complete examples, let alone ones covering long runs of years. Like many, perhaps most, diarists, the farmer seldom gave the full context since he already knew where his property or holding lay, how big it was, and (less certainly) how much capital he possessed. He knew the people he dealt with and who were his workers, and felt no need to write down their names. Sample size is another problem and will no doubt remain so, given the large number of businesses in the farm sector (reported to have been 315,000 in 1851), contrasting with the tiny proportion of accounts ever likely to be found. The sheer number of farmers is what makes it strange that infinitely more research has been done on the other main classes in the agricultural community, landowners and labourers. Everyday decisions by farmers were the ones that pinned the husbandry system together.

Farm record books are private documents. They have the merit of containing definite statements, with little incentive for the farmer to 'doctor' them, unlike present-day versions which have an eye on the taxman and unlike the books of old agricultural writers who were inclined to foist their own notions or nostrums on the reader. The accounts are very often simple cash books, listing receipts and outgoings, say on labour. Mixed with this may be information in almost unbelievable variety. Primarily the records refer to husbandry, the most conspicuous lacuna in agricultural history, where day-to-day farming practice is rarely tackled in close-up. Material is provided on crops and livestock, and innovations can be traced; on borrowing and lending; on output and sometimes on the social life of the farm family. Land use can be deduced and changes in the fields and hedgerows may be mentioned.

This brings us back to the environment. Agriculture was for a long time the largest sector of the economy and what happened on farms is central to understanding environmental change. It might be unreasonable to expect that farm record books would step outside their commercial purposes to detail the effects of husbandry on wildlife but some do. Occasional farmers revealed this as a personal interest.[13] Naturalists' own documents turn up on occasion—Patrick Dillon found a notebook of the leading ornithologist, W. B. Alexander, in the Oxford Oxfam on the day the shop opened and presented it to the university's Edward Grey Institute of Field Ornithology. That was an exceptional find but underscores the point that one should never cease looking. Nor should it be forgotten that estate game books have sometimes been deposited in record offices. Given that shooting was a formative influence in the countryside, these books have from time to time attracted blood sports enthusiasts. In the days when the quarry was regulated only by sporting convention, it can be instructive which species appear in the columns headed 'Various'. The total evidence is voluminous and has not been examined as critically as would be possible. Environmental history is not a small niche to be exploited but a universe waiting to be discovered. Just as we saw that Mauro Ambrosoli demonstrated the viability of a derided example, we may anticipate that his will not be the last revelation of unsuspected meaning in historical sources.

Notes

1. Gillian Tindall, *The Fields Beneath: The History of One London Village* (London: Grenada, 1981), p. 46.
2. Cecil Torr, *Small Talk at Wreyland* (Oxford: Oxford University Press, 1979), II, p. 2.
3. Richard Jefferies, *Hodge and his Masters* (London: Faber & Faber, 1946), pp. 202–203.
4. John Moore, *Portrait of Elmbury* (London: Collins, 1946), p. 68.
5. Michael Bloch, *James Lees-Milne: The Life* (London: John Murray, 2010), p. xiv.
6. Bloch, *Lees-Milne,* p. 95; see also John Harris, *No Voice from the Hall* (London: John Murray, 1998).
7. Sarah Higgins et al., 'Editorial: Estate Archives', *Archives and Records* 40/1 (2019), pp. 1–4 and especially the piece by Michael Moss.
8. Mauro Ambrosoli, *The Wild and the Sown: Botany and Agriculture in Western Europe, 1350–1850* (New York, NY: Cambridge University Press, 1997).
9. W. G. Hoskins, *Old Devon* (London: Pan Books, 1971), p. 45.
10. John Woodforde, *Georgian Houses for All* (London: John Murray, 1985), p. 153, in the chapter on 'Archival Records'.
11. See e.g. Wouter Ronsijn et al. (eds.), *Stocks, Seasons and Sales: Food Supply, Storage and Markets in Europe and the New World* (Turnhout, Belgium: Brepols, 2019).
12. E. L. Jones and E. J. T. Collins, 'The Collection and Analysis of Farm Record Books', *Journal of the Society of Archivists* III/2 (1965), pp. 86–89.
13. One was R. Burder of Rookery Farm, Hurstbourne Tarrant, Hampshire, whose early twentieth-century notebooks I inherited and placed on permanent loan in the Reading collection, and from whose notes the farm's birdlife can be extracted.

CHAPTER 4

Post-war Time Shift

After 1945, when the Second World War ended, seven or eight pre-television years followed. Families still took regular Sunday afternoon walks in the countryside, scooped up arms-full of primroses or bluebells without shame or hindrance, picked blackberries and gathered hazel nuts. Rationing continued. Lacking much access to sugar, children sucked the dabs of nectar out of cowslips as their forefathers had done. Country people still insisted they had a prescriptive right to gather Lapwings' eggs to eat at Eastertide. They climbed up to take Jackdaws' eggs out of their nests in trees and took Moorhens' eggs out of their nests in the ponds. Boys also sought birds' eggs for their collections and it was not really the Wild Birds Protection Act of 1954 that stopped them (although it did usher in a new mood) but rather the novel alternative of television. Egg collecting illustrates the contingency, not just of phases in land use but in the exploiting of nature. Boys had anciently collected eggs, blown them and sometimes suspended the shells from the mantelpiece as ornaments for their mothers, but after the 1870 Education Act, village schoolmasters taught them to build up proto-scientific collections, taking only a single egg from the nest of each species. This was the origin of a hobby taken for granted. Even so, rather than being something that boys had 'always' done, which I assumed was the case because my father had done it before me, this type of egg collecting lasted for barely eighty years.

E. L. Jones, *Landscape History and Rural Society in Southern England*, Palgrave Studies in Economic History,
https://doi.org/10.1007/978-3-030-68616-1_4

The observer's position affects what was noticed. My own notional baseline was not my father's Edwardian boyhood but the 1930s. I was too late to catch more than that decade's fading signs, although twenty years later I did see a man breast-ploughing without horses or tractor in Suffolk and in 1960 talked to a shepherd watching his flock grazing free on Martin Down in South Hampshire, prior to hurdling them at night on the arable fields. The sheep moved about by themselves, accompanied by the clinking of bells around their necks—bells probably made at the great centre of sheep bell making, Aldbourne, Wiltshire. As late as 1961 I saw an old couple haying the roadside at Beckhampton, next to Avebury. Badgers were still occasionally trapped for the sake of eating their hams. Ferreting was far commoner than now. Farmers winked at it because they got rabbit control for nothing. But after the myxomatosis outbreak of 1953–1954, which was the greatest ecological shock of the period, they no longer needed to worry about rabbit control, nor were people keen to go on eating the meat that before the war had kept village families from starving. A family firm of butchers and poulterers in Warminster which had specialised in processing rabbit carcases went out of business. The proprietor rejected sympathy; she told me myxomatosis (commonly known as 'myxy') was the best thing that had ever happened and she turned the premises into a cafe.

One has to think hard to keep the periodicity in mind. Change had been brought to a crawl between 1880 and 1940 by an arable depression which was interrupted only by a brief bout of reclamation during the First World War. My father-in-law went steam-ploughing on Salisbury Plain during that war but the job did not last. Local memoirs report village life at this period or that and it is especially useful to find one by Max Hastings which offers an overview of the years between the wars. His mother told him that as late as the 1920s villages in Southern England, barely forty miles from London, had remained much as they had been in the mid-nineteenth century. Mrs. Hastings bought a cottage at Aldworth, ten miles west of Reading, in 1938, at which date most of the inhabitants still worked on the land. The farms used horses. Grain was cut by reaper-binder. Partridges flourished in the high stubble, cows had horns, the meadows, untouched by chemicals, were rich with September mushrooms. Even fifteen years later little of this would have seemed utterly old-fashioned anywhere in the chalk lands.

Rustic isolation was exceptional but not unique: in 1959 a relative of mine met a very old woman in the canal cottages near Sapperton,

Gloucestershire, who had been up to the village itself only about fifteen times in her life and only once as far as Cirencester, six miles away. Yet change was at length unstoppable. Up to a point Lorna Sage was right up to say in *Bad Blood* that change in the countryside has been ceaseless ever since the enclosures: a commercialised agriculture persists in adopting innovations. Some periods are, however, more sluggish than others. Life had drained out of Southern England during much of the long depression. The most energetic young people moved to the cities or emigrated, leaving behind a landscape of leisure for an elite who spent much of its time and cash on fishing and shooting. Then, almost paradoxically, the Second World War accelerated agricultural innovation while temporarily resurrecting lesser practices that had been ebbing, above all those that helped households to produce a little of their own food. I remember a flitch of bacon hanging in an aunt's house at Hurstbourne Tarrant; the pig-sty, once left empty, had been brought back into wartime use, only to be abandoned again during the 1950s.

Activities that have since dwindled away seemed live enough in those first post-war years. The people who engaged in them thought they were taking up again practices that had been put on hold for the duration. Men from the forces were glad to be home, and while they hoped to change politics, they aimed at the same time to reproduce the daily life they had left at the end of the 1930s. The detritus of war and all the rusting barbed wire was very slowly cleared away; and in some respects the effects and privations of war persisted for years. VE Day and VJ Day celebrated the end of hostilities but marked no sudden breaks in how life was lived.

It may be a little doubtful to nominate any year or even two or three years as the hinge around which 'times shifted', when the effects of underinvestment were repaired, novel goods acquired, old ways given up and new habits adopted. The choice of turning point depends on the indicators selected and there are few unequivocal breaks. Certainly wars stir things up; one has only to recall the cries of John Aubrey and other antiquarians lamenting the break of trend brought about by the Civil War. Nevertheless some habits they might have recognised persisted very late. The land was sleepy and continued so, evolving slowly, especially in times of depression. In 1914 the pubs were rustic and unspoiled, at least in the Cotswolds, as the Arts & Crafts architect, Norman Jewson, described them in his privately published autobiography, *By Chance I Did Rove*. Many inns, he said, had been unaltered for a century.

In the 1920s, the roads were still white, not black, there were no telegraph poles, and cottage gardens were bright with flowers. Rural poverty was all too evident; village children might go without shoes and when my father was installing wind-pumps on Salisbury Plain in 1922, he was scandalised to find the children of the farmers themselves going shoeless. Before the First World War women in Longparish had gone to their market town, Andover, only twice a year. It was four miles away. But after the war, surplus lorries were converted into charabancs and then into country buses, so the women went regularly and even began to work in the town. For all that, Jewson exaggerates when he says that the First World War changed 'everything'. There was little money between the wars to experiment. My father told me that the big chalk land farmers I knew as prosperous in the 1950s used to come into town in the 1930s with their cars tied up in binder twine. The difference was that farming was no longer depressed after the Second World War, an upheaval which was followed by rural society's biggest shift, assimilating the countryside to mechanised agriculture and urban lifestyles.

About 1950 agriculture was really intensifying. Tractors displaced horses. New machinery made reclamation easier. Combined harvesters were multiplying. Herbicides and chemical fertilisers were fast being adopted, milking parlours replaced cowsheds, and round grain-storage tanks sprouted in the old rick-yards, and silage-making started. Haymaking ceased to mean pitching loose hay and soon the vast majority was baled. All these developments came in a rush within three or four years at the very end of the 1940s and start of the 1950s. They were the background to the conversion of most of the remaining down-land to arable fields. Meanwhile the decline in the number of farm horses impoverished those blacksmiths who could not make a living from fabricating ornamental ironwork and brought the trade of wheelwright to an end. Coach builders shifted to making motor cars. (As late as the 1970s the Morris motor company used wood in one model, as if a vehicle for the mass market could really be what it was then called, a shooting brake.) Incomes were rising and a new form of leisure, bird-watching, joined the traditional hunting, shooting and fishing. The era of mass bird-watching began symbolically in 1955 with the arrival of coachloads of people to see bee-eaters nesting in Sussex and the next year to see hoopoes nesting in Hampshire.

The post-war years, 1953–1954, can reasonably be taken as the fulcrum of change. Not coincidentally those years experienced the myxomatosis plague. A whole economy had rested on the rabbit. Skins were collected by dealers and it was permissible to send carcasses by train, harled but not wrapped, with only an address label tied on. Rabbit meat had been what had enabled country families not only to get through the depression but through wartime and post-war rationing too. Now it vanished. Down-land ecosystems were transformed, because the removal of the rabbits meant the short nibbled turf grew up, choked the orchids, reduced short-grass species of butterflies and started a slow take over by scrub.

During the first half of the 1950s many oak-hazel woods were still being coppiced, creating glades suitable for light-tolerant species of butterfly. Sheep folded in pens to manure the land were ceasing to be profitable and the demand for hurdles to make the pens evaporated. Even today hazel stools remain everywhere but they are grown out. Like the ash trees discoverable along the bottom of hedgerows, bent over by the last man to lay the hedge, they persist as reminders of old ways in the present day when hedges are controlled by crude flailing. If the woods are now used at all they are where gamekeepers house pheasant poults ready for the shooting season or park a few feed bins. In small copses there is little woodland management and close-packed saplings strain upwards in competition for the light.

My boyhood woodland haunt was Bilgrove Copse, then called Belcher's Wood, which was small and mixed and more to the point an easy cycle ride from Andover. It held elms and a clump of beeches where rooks roosted in spring and chaffinches foraged for beech-mast in autumn, but was mostly plain North Hampshire oak-hazel coppice. The oak standards were not large and had been planted at intervals in a sea of hazels which at the start of the 1950s were still properly coppiced. In Februarys two woodmen worked cutting and piling up hazel wands ready to make hurdles. Wood chips, twigs and small branches were got rid of in a fire that was always alight somewhere in the clearing. The woods were quiet—chainsaws were on history's horizon but had not quite arrived. Charting their take-over is oddly difficult, the story being written by enthusiasts for particular brands of machinery who have no wider interests. In the early 1950s I had never heard a chain-saw and expected the loudest sound in the woods to be a bang when someone potted a grey squirrel.

Coppicing meant that at any one time a sizeable patch of wood was left open while the cleared hazel grew up again from the stools or clumps. The next season the woodmen moved on. With light getting into the clearings, primroses and bluebells could spring into life. Butterflies could thrive. Where had they been when it was all dark cover? Furthermore, a cheerful farmworker spent part of January cutting the hazels along the main ride and heaping up branches whose young catkins shook sorrowfully as they died in the breeze. The man's work added to the area open to the sky and was meant to provide his employer with a field of fire wide enough for casual pheasant shooting. All this was very Victorian; coppicing is long over now. I had been watching a late example of traditional oak-hazel management.

The compartments of the wood turned shadowy and dank when the hazel was re-growing, as it would be for seven or fourteen years until it was cut again. (I left home finally in 1958 before ever seeing a full rotation but it was probably never completed.) Even in bright summer, leaves would block out the sky in the deep woods. The grown hazel woods were strangely lifeless. Coppicing and cutting saved the place for the naturalist. The modern fetish of trees for trees' sake had not taken over. I could glance around in the old days and spot the tit flocks of autumn as they arrived at the edge of the clearings. More important was what the open stretches did for butterflies. The ordinary species of the countryside were present, together with a healthy population of silver-washed fritillaries, white admirals and commas. Once, while approaching along Hungerford Lane, I found a white-letter hairstreak, rare enough nowadays to be exciting and not often seen in my boyhood. The species lived at the tops of elm trees. Since the epidemic of Dutch elm disease in the 1970s, elm trees have disappeared but white-letter hairstreaks have managed to adapt. They now survive obscurely on the suckers of the English elm, which die off after a certain height, but live most successfully on wych elm. As canopy feeders they are neck-achingly hard to find.

Coppicing was not the only time-honoured practice shrinking fast in the first years after the war. When springtime weather had its liberating effect in April and May, Belcher's Wood was infested by couples, families and quite large gaggles of people, all picking primroses, bluebells and catkins on weekends. In the early 1950s flower-picking was normal, yet, as the last phase of a pastime dating from when country people had few aesthetic outlets, it was nearing its end. Public opinion now frowns on the practice. Even going for sedate country walks on Sunday afternoons, as families once did, has become unusual. After a few years this withdrawal from the countryside was to become evident as the old droves had fewer feet to tread them down (and no sheep to be driven along them); they narrowed and the hedges grew in from both sides. The drove between Andover and Winchester was wide enough in 1944 for my grandfather and I to find Canadian trucks backed end-on into the hedges as they prepared for D-Day. It narrowed in ten or fifteen post-war seasons to little more than a footpath.

In reality coppicing was in steep decline and the farms were busy mechanising. I do not recall horse agriculture but do remember going with my father in 1952 to a presentation about the new Ferguson tractor, more nimble and versatile than any other available. Compared with modern farm machinery the little grey Fergie seems a toy. In my wife's village, a London estate agent's wife bought a farmer's redundant plough horses and hired a man to look after them, because she could not bear to see them slaughtered. She may as well have been King Canute. By then the gamekeepers had been demobbed and were starting to shoot and trap with their old abandon. Their gibbets of dead stoats and sparrowhawks, crows and magpies, could be seen hanging in full view at the edge of any copse so that their employers could see they were doing their job. Pot-hunters wandered openly with guns, hoping to take what they could from the land, as they had during wartime and rationing. Within a very few years they were stopped by the keepers. Freedom to roam lasted at most a decade.

The Wild Birds Protection Act of 1954 did gradually make a difference. The police inspector to whom my father and I gave a lift after church snorted when I mentioned the Act but did set a young constable, armed with the little Observers' Book of Birds, the fruitless task of catching offenders. Television did more to coax small boys out of the hedgerows than Parliament or the constabulary ever could, but the Act crystallised a shift in public opinion. Here, however, I am speaking of the eve of change when television sets were uncommon—people crowded into some fortunate neighbour's house to watch the Coronation in June 1953. Until then egg collecting was what one imagined boys were up to if they were found fossicking about in the nesting season. Even so it was possible to mistake their aims and in Belcher's Wood I started to bawl out a gang of youngsters climbing trees until they showed me they were taking only pigeons' eggs, for which they had permission. Embarrassingly, two of them proved to be connected to me by semi-imaginary chains of second cousinage.

One was the son of Sonner Bonner, thatcher and landlord of the Pig & Whistle (Hare & Hounds), which lay along Hungerford Lane, not far outside the wood. Sonner was described by his wife as a moucher, an opportunist poacher. He was always slipping out at night and was adept at guddling trout and taking deer, not that he let himself be caught doing that close to home.

In Belcher's Wood more shots were fired at grey squirrels than at everything else together. Squirrels were detested because they took pheasants' eggs and harmed young trees by stripping the bark. They damaged the hazels and scavenged the nuts. Their tails attracted a bounty of one shilling. Small wonder, the farmer and his son seized every chance of taking pot-shots at them. After feeding their few pheasants they would walk along with an empty bucket on one arm and a gun under the other, when, clang, down would go the bucket, up would go the gun, and bang, the next thing one saw was the ejecting of the cartridge. As pest control the shooting was fruitless, for the wood stayed alive with squirrels. In February and March gin traps were set for them in little triangular hutches baited with grain but, although I have seen twenty lain out dead in a row, the numbers did not seem to be dented. Chaffinches, meanwhile, prospered on the spilled grain.

I had come to Belcher's Wood to watch birds, not to observe the ways of the countryside. For the time being, wild birds were my all and rural life was a backcloth. Like most young people I assumed the manners and pastimes around me to be timeless. Had I thought, this would have seemed the nonsense it always is and I would have taken more notice of village habits and farming practice, but I anticipated nothing beyond a world of rationing, limited travel, restricted communications, and rural people seizing such few gifts as the heavily privatised land offered. The idea that the early 1950s were the absolute last gasp of old long-standing ways did not then occur to me.

PART II

Localities

CHAPTER 5

Chalk Downs

The history of the Wessex chalk lands is not what might be expected and demonstrates the element of contingency in land use. The point can be brought home straightaway via a preamble about my own early experiences. Start, then, with the fact that the best of the places accessible to me as a young naturalist was a down-land valley hidden in waterless, hilly country to the north of Andover. Its contours on the marvellous one-inch Ordnance Survey map came together like a concentration of isobars on a weatherman's chart. Thinking to explore the possibilities I rode out one day in the 1950s, hid my bike in the hedge, crossed a stubble field and came abruptly to the top of a yawning dry valley. The turf on the slope was covered with flints and rabbits, more rabbits than I had ever seen. The whole hillside moved with brown fur.

No one was about and an old keeper's house that backed into a bit of woodland on the far side was long deserted. The landscape seemed to be a relic of agricultural use that had ceased. I went gingerly down the steep hillside and looked around. The valley was U-shaped and could be followed until rising ground at the far end of the horseshoe blocked it. There were exciting birds to be seen, for this was genuine old-style, short-grass down with the species that must have typified the high chalk between the wars and perhaps far longer, since little of the ground appeared to have been ploughed in recent centuries (prehistory

E. L. Jones, *Landscape History and Rural Society in Southern England*, Palgrave Studies in Economic History,
https://doi.org/10.1007/978-3-030-68616-1_5

was perhaps another matter). It is pleasing to find that I have kept eight or nine photographs of the valley as it was in 1953, mere black-and-white snapshots but fully indicative of the topography and vegetation. About that date I took a few photographs of other down-land habitats in the Andover district which also capture the last of the pre-war depression, small untended spots, excellent for wildlife. In 1961, when I was a research student in agricultural history and had realised how fast the older countryside was vanishing, I took about 200 photographs of habitats, farm buildings and associated items. Many were of the Lambourn Downs where remnants of chalk grassland survived, although they were on the retreat. The photographs included stooks of oats and a binder, thatched barns and proper straw ricks. Today the scenes look as old as the hills.

At some stage, fields in my secret Hampshire valley had been divided by a hedge but in 1953 the scene showed few signs of recent activity. The slopes were too abrupt to plough with heavy modern machinery. They were dotted with scrub and at their foot was a line of well-grown elder bushes that looked like volunteers. Current occupation gave the impression of being limited to the desultory grazing of sheep and some game shooting. Wheatears flitted about the patches of flint and bare chalk, redstarts showed their tails in the elders, woodlarks bounced away and a stone curlew lifted off with an unexpected array of black and white on its wings. Fallow deer looked up and were off. Turtle doves were purring in the wood; but, rare now, they were then nothing special. On later visits I added buzzard, hobby and Montagu's harrier, and was drawn to a winter great grey shrike when it chattered at two crows flying low overhead. A friend and I found a dead fox with a mangled front paw and a broken back leg, obviously having been trapped. That was what warned us to take care when we came across a square of smooth earth whose turf had been skimmed off. In the centre was a little hillock with the remains of a rabbit and a buzzard's pellet. I stepped off the patch and a gin trap leapt up through the soil and snapped shut. Poking at the ground I found another which sprang up and caught the stick. There was a total of four gins and beneath the dead rabbit a fifth which nearly caught my hand when I picked up the pellet.

Another day there was a green woodpecker on the grass, fluttering but not flying off when approached. It was trapped in a gin and I had to kill it. Full of indignation, I reported it at a village police house on my way home. The bobby outrageously carried my report to the farmer, who turned out to be the father of people with whom I was at school. That ought to have put me off visiting the valley (and thinking the police force of rural Hampshire would be impartial) but birds just too good to miss drew me back. No equivalent habitat existed in the immediate district, although there were still one or two unploughed downs further north on the high chalk near Walbury Hill. Salisbury Plain, only a few miles west, could have matched the birdlife but the army kept it out of bounds. I did not have a first view of the sweeping grasslands of the Plain until the early 1960s, when the village of Imber was briefly opened to a public protest against the failure of the military, which had seized it for wartime training, to keep its promise to let the inhabitants return. A long convoy of cars belonging to 'activists' queued up but we drove right past, intent on seeing the sights rather than making a political point.

Military land is now more accessible than it was but private land is less so, thanks to the craze for shooting. More intensive farming and the progressive destruction of down-land remnants make the chalk ornithologically less appealing than in the middle of what must now be called the last century. Few birders are concerned; they prefer wetland habitats and in Hampshire that means the coast. They rush off after every report of an unusual bird. Watching one's local patch, correctly described as a 1950s' practice in Bill Oddie's *Little Black Bird Book,* has given way to chasing rarities wherever they appear. As to the rewarding down-land valley which had become one of my haunts, the gamekeeper eventually caught me. In the way of his kind, he would not even tell me his employer's name so that I could write for permission. But the date was September 1955, I was about to go up to university, and let the matter rest.

In 1983 two economists began a book on *Agriculture, the Countryside and Land Use* with a regretful passage about the destruction of Danebury Down, south of Andover. They first wrote the piece in 1969 and I supplied the information, Danebury having been another resort of mine during my schooldays. It was one stop on a round trip by bicycle which, in various combinations, took in the confluence of the rivers Test and Anton, Chilbolton Common and West Down, Brockley Warren (out on the downs towards Winchester) and the only open water in the district, at Marsh Court, just down the Test from Stockbridge. From the purely ornithological point of view all these places were too precious to lose. Yet in the 1950s and 1960s several were already going. Compared with many districts, especially on the coast, they produced rather few species and especially few of the stray drop-ins that give spice to birding. They had been the best of a restricted hand of cards. To change the metaphor, the palette was limited but one had to paint a picture as best one could.

The book on agricultural economics described Danebury Down as a classic chalk land habitat of unsown pasture interspersed with hawthorn scrub. Above it, surmounted by a beech clump, was what was regarded as an outstanding example of an Iron Age hillfort. On the lower, more level, ground lay the abandoned grandstand of a racecourse that had been famous in the nineteenth century. In the 1950s a now almost unimaginable richness of birdlife existed here: red-backed shrikes, woodlarks, redstarts, hobbies and long-eared owls. The flowers were typical of good chalk grassland, including several orchids. By the junction of the road and tracks where I always parked my bicycle was a fringe of yew trees—the 'Hampshire weed'—from which in winter short-eared owls wafted away ghostlike as I walked through.

Even in the 1950s few spots like this remained and in 1960 Danebury too was ripped up to grow crops, despite having escaped the ploughing of the war years. Agricultural subsidies made cultivation worthwhile for the farmer, although scarcely worthwhile for society because food could have been imported more cheaply than grown at home. All that lasted after the new ploughing was the prehistoric encampment, a landmark destined to become the site of a famous archaeological excavation. The county council bought it. Archaeology had prestige; ecology and environmental history had none and the down proper was let go under crops. At first the council omitted to ensure public access to the hill and a reporter from the local paper went in by helicopter to dramatise their oversight. Undoubtedly the farmer held them to ransom. Like too many big farmers he was a nasty man to meet. The farmers of North-west Hampshire had been castigated by Colonel Hawker of Longparish one hundred and thirty years earlier and had not changed their spots. I lamented the destruction of the down and approved the economists' label of classic chalk land habitat. If truth be known, I probably supplied the term. The full story about land use on the chalk is however more complicated than the phrase suggests, as becomes clear from the fate of a lesser down, West Down, Chilbolton.

For over a century before my birth the rural economy in which I grew up had been depressed. One of the early topographers of Hampshire got it about right. Most such writers tack from place to place noting the main features, especially the churches, and dispense praise and blame without analysing the forces bearing on the landscape. Robert Mudie, though, did grasp the changing economic context in his three-volume

History of Hampshire of 1838. He was a Scot who worked in Winchester and his outsider status may have enabled him to see further than others. Germaine Greer dismisses him as 'a desperate hack' and it is true he wrote, or compiled, ninety books in his lifetime. Why did she think trying to make a living by the pen was so wrong? Mudie was more astute than she suggests. In his section on the Test Valley he introduces the division of labour in so many words in order to explain what was happening on the local scene. The phrase 'division of labour' is of course attributable to a fellow Scot! Mudie noted that industry had shrunk away in the towns of Southern England, which he thought was due to northern competition, and although that was only part of the story, northern industry doubtless stifled any resurgence in the south country. He recommended that the district get on quietly by concentrating instead on agriculture.

What Mudie could not have foreseen was that within little more than another generation the arable farming of the downs would in turn be undermined, this time by imports, and plunged into the Great Depression. The living standards of consumers rose as imported wheat made for a cheaper loaf but domestic agricultural production took a knock, just as manufacturing had done before. Investment in farming fell, there was less work in the cereal-growing districts, the more aspiring men moved away. Cottage England was lovely if one did not look too closely or have to dwell in what, realistically speaking, were rural slums. Ironically, North west Hampshire's beauty was often the consequence of decay.

There were a few counter movements. Between the wars occasional firms had arrived, such as Kelly's Directories in Andover, but the market towns were low on industry and conspicuously light on both public and private investment. Then the war and post-war years put money back into the pockets of farmers and landowners. For the moment, change hung fire. In the 1950s there were still country cottages abandoned in the depression and crumbling away where today any shack qualifies for conversion to a second home. The only upturn had been when the First World War pumped in military money but little of that had carried over. The Second War brought larger (and because of the Cold War more lasting) investment in the form of defence establishments. A number of the wartime camps were not however needed; they were abandoned after 1945, littering the district with derelict buildings and rusting barbed wire that took years to clear away. Numbers of military officers retired to the local villages, it was said with some exaggeration because of the fly-fishing. The culture they introduced by clinging to their wartime ranks added another ponderous layer to the dominance of landownership and farming. Jonathan Meades' *An Encyclopaedia of Myself* has them 'off to a T'; his father had been a wartime officer.

The large bomb dump in Harewood Forest near Andover was an example of hangover militarism that took ten years or more to clear. There were plenty of other disfigurements, a minor one being a concrete pillbox sited at the corner of West Down, Chilbolton, commanding the road and railway bridge near the confluence of the Test and Anton. The housing for a gun, possibly only a bren gun, was still there in the mid-1950s and could be swivelled in make-believe resistance to Huns advancing up the valley from the south coast. Higher up the slope was a rubbish tip where I once saw a couple picking through the refuse for anything saleable, while below a fly-fisherman was casting on the expensive waters of the Test.

This vignette of English inequality ought to dispel fond memory. But for me West Down must eternally be bathed in a romantic glow. There were red-backed shrikes in the hawthorns and nightjars churring in the dusk, stone curlew skirling from the old airfield, lapwing and redshank calling from that direction too, and a range of other water birds active along the rivers and in the wet meadows of the valley. West Down itself was an irregular slope of grassy patches and hawthorn scrub, full of delights for the naturalist (and lover) and no longer in agricultural use. A farmworker called it 'fuzz land', meaning furze or gorse. Far enough from the village to dissuade even country dog-walkers, almost no one disturbed it. In the 1950s its special birds, the shrikes and nightjars, were not yet special enough to draw in bird-watchers, of whom there were anyhow vanishingly few in the Andover district. It was an English unofficial rose, to be plucked and tossed away in the following fashion.

One evening during the Easter vacation of 1957, when a bitter nor'-easter had subsided, I went out to where the Anton joined the Test to see whether swallows and martins might be migrating through. Leaving my motorbike in the bushes, I walked up the edge of West Down towards the airfield. The unexpected sound of an engine grew louder and louder so I turned towards it onto the Down, treading the springy turf that was one of its pleasures. Unfamiliar brown patches came into view and I burst suddenly into a raw empty space. The hawthorns had vanished. The ground was churned up, with the turf removed and odd branches and roots scattered about. In the middle distance of this destruction two or three of the bigger trees did stand upright, their trunks battered and scarred and with young green foliage and bonfires of branches from uprooted hawthorns heaped at their feet. Beyond was a retreating line of bushy trees, darkening against the sky, with a giant yellow bulldozer backing away.

The driver seemed glad to stop. He switched off and pulled out his Woodbines for a smoke. He could tell me little about the job because he was just a contractor's man glad of working for overtime. He admitted the work was not hard, as long as one watched out for branches hitting one's eyes. There were no tractor cabs in those days. The task was being undertaken for the Air Ministry, a representative of which had arrived that morning and graciously commanded him to 'leave a few trees'. A block of eighty acres was to be cleared and seventy-four of them handed to 'the agricultural people'. Redshanks started tootling on the airfield. Darkness was falling, in more ways than one.

The material point is that what I had been seeing in the remnants of down-land grass like West Down was land vacated by agriculture. By all accounts many parts of the landscape had become progressively more wooded ever since the retreat of the nineteenth century's intensive exploitation, but this particular down had not been overwhelmed. The surviving grass tended to be in patches which, because of steep slopes and limited supplies of fertiliser, had been physically difficult and too expensive to bring into cultivation during the wartime ploughing campaign, although the bulk of such land had succumbed. During the 1950s, subsidies and better equipment started an assault on almost everywhere that was left. Places that so far had escaped the plough were subtly reverting to coarse grass and spreading scrub. They were already not as appealing to the naturalist as they had been.

I was heart-broken when the Down was bulldozed and the land turned into a chicken run. Nevertheless my view did not make full ecological sense, which is the essential lesson about the history of land use. I had not been around long enough to realise the contingent character of what had been there and was now being obliterated. By this I mean not only was I a child of my (then short) time but that West Down itself in all its glory was a passing creation. It had been evolving so slowly as to appear a fixture. Conservationists, had the district then boasted any, would have wished to keep it in its 1950s condition, as later ones definitely did. That would have meant arresting the plant succession and holding it for ever in suspended animation. The task would have been costly: the scrub would have had to be kept pruned and a way found to keep sweet a sward already starting to coarsen now that myxomatosis had despatched the nibbling rabbits.

A little more discussion of the issues will make clear the impermanent nature of what I was lamenting. Ecosystems can be understood only in terms of their historical period and as artefacts of the economic forces moulding them. The present is not a static hand-me-down, eternal as each moment must seem to the individual naturalist. The phases of agricultural history have always to be taken into account because the price incentives that lead to the growing of one type of crop and creating one type of landscape rather than another vary from each period to the next. What makes this hard to recognise is that a given price regime may last longer than the lifespan of a single observer, though not necessarily much longer.

Naturalists will find this approach unfamiliar and even uncongenial, as I would have done in the 1950s. But the interpretive effort has to be made. West Down was only a remnant. It was not truly in equilibrium nor was it an inert fossil. The forces acting on it were bringing about changes which within a couple of decades must have suffocated much of the existing plant- and birdlife. Myxomatosis brought the greatest change but unregulated plant growth was already rearranging things. The Down had not long been in the state I witnessed, since the evocative ecosystem of abandoned agriculture in the mid-twentieth century could not have been afforded under the farming regime of the eighteenth and nineteenth centuries. Land in earlier centuries was needed for producing food and other resources. What was there previously was sheep-down used to supply food, not a languishing mixture of scrub and patches of grass attractive to nesting birds.

Old Shepherd Vincent had been the patriarch of a family whom I knew in Chilbolton. He told the author of *The Common Lands of Hampshire* what West Down had been like in the early twentieth century. It was divided into three strips belonging to different farms. The parcels were marked by boundary stones but not fenced or hedged. The sheep were given the run of the whole and if the flocks got a little mixed they sorted themselves out by nightfall. They were then led off to one or other of the fields on their owner's farm and penned within hurdles to manure the soil overnight. The entire ecological system fitted neatly together, with a regular coppicing of parts of Harewood Forest and Wherwell Copse to supply hazel wands for the hurdles. In spring the water meadows of the valley floor were 'floated' (drowned and dried off alternately). The flocks were moved there to take advantage of the fresh growth of grass in the

April 'hungry gap', when winter stocks of feed were used up but the grass of dry pasture would not yet have appeared. In the mid-twentieth century the ridge-and-furrow, banks and channels, of these meadows remained as archaeological features on which Redshank, Snipe and Yellow Wagtails nested. No longer, the dog-walkers have driven them off.

After each spring the sheep were again grazed on West Down. Its turf was kept short and intrusive seedlings were nibbled away. The birds I saw there had probably colonised the Down from gorse brakes that had interrupted an open hill slope, because gorse had its own value. There is no reason to assume that the birdlife had always been as varied and numerous as it had become by the 1950s. A vision of the late nineteenth and early twentieth century as ecologically bountiful colours our view of the past. Farming in decline and collector-naturalists on the ascendancy make those years seem magical and mislead us into thinking they had been the ever-present norm. The sheer abundance of butterflies celebrated by Victorian entomologists has encouraged a mordant declinism in which populations are believed to have been always larger and habitats always richer at any given date than at any subsequent time. But before 1880 the need for food and the rural habit of using every resource at hand had precluded a casual use of the land. It was the depression of farming that relaxed the pressure and gave space for wildlife.

The grazing of sheep on West Down ceased between the wars but the village clung to the idea of access. A Mrs. Disraeli, daughter-in-law or some such relative of Benjamin Disraeli, lived in a big house by the river and before the Second World War had 'Private' notices put up. The local men threw them down and she did not persist. The military commandeered the place in wartime, built the corner pillbox and some hutments near the airfield, but had no real use for the grassy slope. Rabbits took over. They kept the sward sweet but did not hold back the advance of scrub as effectively as sheep had done. When myxomatosis hit the rabbits in 1953 and 1954, hawthorn and woody shrubs started to creep forwards. Wildlife found a patchwork of grass and scrub to occupy. I was enjoying an interlude that was not ageless, was not going to last, and has not lasted.

No official body was prepared to put in the energy needed to manage the Down for the sake of wild flowers and wild birds. In the 1950s none would have seen the point. The authorities never did. As Eleanor Lockyer details in a privately published history of Chilbolton, the rector had led a protest against building on Chilbolton Common as early as 1890. In 1957 the Air Ministry sold West Down to Andover Rural

District Council, which started to dispose of parts and let the remainder become a wasteland; land allotted to the poor at the enclosure in 1858 was enclosed in 1963 and despite a struggle only a little has been recovered; the merest glimpse of the river Test from the riverside path was blocked by a very high fence; and the ancient drove called the Mark Way was closed, although after a fight a new footpath was established along it. West Down has a history emblematic of the lack of concern which the public authorities show towards public open spaces.

Turning West Down into a home for battery hens during the 1950s might be thought justified on grounds of the private property rights that underlie the whole economic system. An owner ought to be permitted to make what use of his land he may choose. I shall not argue the case either way because, for one thing, I am not sure who really owned the down. The community held at least prescriptive rights. In Victorian times the gorse had been cut and bundled for firing by the shepherds and other villagers, an annual practice that seems to have been carried out on some scale. Gorse was valuable for bakers' ovens and for the firing that was always scarce on the chalk hills. At the start of the war the Air Ministry took over the area and may simply never have handed it back. Talk of the parish acquiring it as a leisure resource (an anachronistic concept anyhow) before the chicken run was built remained councillors' chatter. There were similar cases locally, Hurstbourne Common to the north of Andover being a notorious example; an uncle of mine was the last man who bothered to exercise the right of cutting pea- and bean sticks there. In the early twenty-first century the military authorities have been advertising in the farming press for anyone to come forward whose land, or whose family's land, had been seized for airfields and camps. I guess more defence land was finally to be sold off and the aim was to draw the teeth from any rear-guard claims. The case for preserving West Down, Hurstbourne Common and anywhere else I know in the district was never pressed.

A better argument than the sanctity of property rights might be the effect on the national food supply. After all, the role of chicken in Britain's social revolution can scarcely be exaggerated. What statistics does one want? A mere one million chicken were consumed in 1950, when poultry was a luxury eaten at best on Sundays and by many people only on high days and holidays, yet by 1967 annual consumption had risen to 200 million. Whereas in 1950 poultry made up one per cent of all meat consumed, it accounted for 25 per cent by the 1980s and had become the single most important type of meat. This was the result of technological changes making possible more intensive rearing. The comparatively small farm on West Down played a part in this culinary revolution and it is not clear that from society's point of view preserving a bit of scrubby down-land edge would have been a more desirable use.

This might however develop into an argument for converting each and every scrap of unused land to agricultural use—an aim which, if genuinely urgent, might surely have been furthered earlier, during the war or the dollar crisis, when food was truly scarce. Why had West Down been left idle until the late 1950s and only then destroyed? Subsidies and the overweening political influence of the farm lobby are the explanation. The apparatus of conservation was weak, witness the failure to protect sites of special scientific interest (SSSIs) from the farmers who in the 1960s ploughed up hundreds of them with the same abandon they had been expending on levelling prehistoric barrows ever since Richard Jefferies complained about this in 1870. The designation of SSSI was intended to preserve biologically important places. Unfortunately it was carried out in a very low-key 'English' fashion, like the current voluntary and amateur listing of public rights of way—inefficiently and nothing to celebrate.

The catalogue of sites of scientific interest was drawn up via casual listing by prominent naturalists—for instance the then Nature Conservancy admitted they knew little about Yarnton Mead, Oxford, an ecologically exceptional site to be discussed in another chapter. They told me curtly that I was likely to know more about it than they did! Lesser places without unique features were not recognised, not listed and received no protection. Their only chance of survival was if they were known to be prime archaeological sites or more likely someone's pheasant shoot, just as its role as the local MP's shoot rescued Knighton Bushes near Lambourn from becoming a bombing range. The ambitions of the military were unbridled; Stonehenge itself had only narrowly escaped demolition in favour of a flight path. Although West Down possessed a little archaeological significance it did not have enough to qualify and the patch had somehow escaped the grasp of the game preservers.

Such places were emphatically not changeless grassland that had been peppered throughout time with orchids and inhabited by red-backed shrikes, redstarts, wheatears, woodlarks, nightjars and stone curlew. By 1961 I had come to comprehend the historical, or dynamic, nature of chalk down-land and wrote a letter to *The Times* (7th September 1961) suggesting that conservationists should run sheep flocks on the grass to keep back the scrub. I wanted to see experiments with folded flocks, old-style, to simulate their earlier use: the notion of living historical farms was however simply not understood in this country and the fact that the concept was American would probably have damned it. Max Nicholson from the Nature Conservancy came to see me in Oxford but I had just changed colleges and we missed one another. Years passed before environmentalists actually began introducing sheep (and in river meadows, cattle) to control the growth of vegetation.

Although historical records resemble only a spotty and blurry film, they are consistent with the long-period alternations between pasture and ploughland discernible in the writings of John Aubrey in seventeenth-century Wiltshire, Gilbert White in eighteenth-century Hampshire, those of the Hewetts, father and son, in nineteenth-century Berkshire, and forward to any number of early twentieth-century commentators. The details are complicated and 'post-dicting' fauna is difficult because not all species respond to agricultural change equally energetically or in the same direction. Nevertheless, at the stage when a given type of land use was dominant, successive writings show that it supported approximately the expected birdlife.

The long-term swings and roundabouts of land-use on the chalk downs reveal how arbitrary conservation must be. There is no unambiguous baseline from which to measure change: 'you pays your money and takes your choice'. The question is do we prefer arable land with flocks of linnets, common at the present day, or hanker after the old grassland with woodlarks which are now uncommon? This is the trade-off. Admittedly both outcomes are available in principle. Unfortunately in the current arable environment maintaining chalk grassland on a grand enough scale to sustain even a simulacrum of the old birdlife would be expensive. Former ecosystems do remain on some scale because the military once had the money to secure Salisbury Plain as a great unintended nature reserve. How stable their ownership will prove may soon be seen. Yet the choice between types of land use does not rest on scientific evaluation but on subjective preferences and value judgements. Linnets or woodlarks? It is easier to spot the dilemma when the history of fluctuations is recognised.

CHAPTER 6

Heathland

Encroachments in the New Forest

One evening in late spring, 1961, I arrived from Oxford at Colin Tubbs's flat in Lyndhurst. Colin was the Nature Conservancy's warden for the New Forest. The next morning he took me onto the open Forest, stopped the Land Rover alongside a heather moor and gestured towards a large patch of gorse a couple of hundred yards away. 'Look at the edge', he said. We got out and clambered onto the bonnet. From that slight elevation the block of gorse was obviously straight-sided and when we walked over to it we found a distinct bank-and-ditch along the outsides of a rectangle. 'One or two more like this', he remarked laconically, and drove to them. Keyed up to find something more than formless masses of spiky green gorse bushes with occasional yellow flowers, which was all anyone had noticed until then, we spent a number of weekends finding a whole set of regular forms.

Once the idea strikes that a landscape feature is distributed systematically rather than randomly—provided the idea is at least half right—further examples fall into one's lap. This is what happened to W. G. Hoskins in the 1940s when he noticed the clustering of late sixteenth- and early seventeenth-century dates on a handful of Leicestershire manor houses. He looked for published confirmation, found examples of the same period tumbling from the pages of one unregarded village history

E. L. Jones, *Landscape History and Rural Society in Southern England*, Palgrave Studies in Economic History,
https://doi.org/10.1007/978-3-030-68616-1_6

after another, where local authors had noted the date of a house but had not joined the dots because they never strayed from their own parish. Hoskins realised he had discovered the phase of architectural history which he labelled the Great Rebuilding. A similar Eureka moment struck us in the New Forest when we noticed the repeated correspondence of gorse brakes and sites of former cultivation.

Most of the gorse was on once-disturbed soil. A little of it straggled along the edges of tracks. Some were in patches that did not easily reveal their geometric shapes if the eye failed to pick out the shallow ditches and overgrown, enclosing banks. The bulk proved to be on low banks surrounding enclosures of up to a couple of hundred acres in size, always with an outside ditch. These enclosures simply should not have existed on the open Forest, which was Crown Land; as a result archaeologists did not believe they existed. We explored most of the Forest and started to look outside the Perambulation, as the legal boundary of the Forest is called, because I had come across an old publication describing an intake of 1805 on a common towards Christchurch.[1] The diagram and map in the article could be followed on the ground, though whatever function the enclosure may once have performed had long ceased. The site blurred into heath and Forest-edge fields.

We found twenty or thirty unrecorded sites and were delighted by the correlation between the distribution of gorse and human interference with the soil. We became enthusiastic about using gorse as an indicator of archaeological sites, though puzzled and frustrated that the enclosures and earthworks, not to mention the gorse brakes themselves, seemed undateable. No potsherds or other fragments lay on the ground, not even in the entrances. Maybe it was just as well. Speaking of the New Forest in his *An Introduction to Field Archaeology as illustrated by Hampshire,* Williams-Freeman remarked that, 'there is an old man in the neighbourhood, who says he used, in his youth, to put bits of pottery into the hill in order to get the employment of digging them out!'

No pottery taxed our brains at the gorse intakes; there were no traces suggestive of settlement. The banks and ditches were in varying states of repair. Some were sharp, as if they had been built up or dug out only a few years before, although no one remembered them in use. Others looked low, worn and ancient. We could separate enclosures into different types by shape, size and location—make a typology—without being able to assign a date to any of those on the Forest proper. No modern maps recorded them and the archaeological literature contained no mention.

Within their bounds, several of the enclosures bore the grass- or heather-covered indentations of old ploughing rig but others were barren of markings, as if they had perhaps been stock pounds. Commoners lived as they still do in cottages and farmed a few acres along the fringes of settlement. Their right to run livestock on the forest gave them extra grazing and made the effective size of their holdings bigger than the little paddocks around their cottages might suggest, but they were still as close as the English came to being peasants. They were the least likely among farmers to have kept records and none turned up. There was little private documentation of any kind, which was not surprising if the enclosures were illegal intakes on Crown land, conceivably winked at in times of stress and high prices but afterwards abandoned or thrown down. This was turning out to be more archaeology than history, and baffling at that.

A local antiquarian, Heywood Sumner, had identified 'bee gardens' on the fringes of some New Forest heaths. They were small enclosures that had supposedly protected beehives from ponies and cattle. The enclosures we were finding were bigger, up to 200 acres; even the lesser ones were too large to have been constructed surreptitiously. There had certainly been a saga of 'rolling hedges', when commoners had pushed out the boundaries of their gardens and paddocks, only to have them spotted by the keepers and thrown down, but nowhere was there any record of identifiably separate fields or holding pens out on the heaths. Given the supposedly inviolate status of the Crown Lands, the scale and number of enclosures were astonishing. Gorse was proving to be an anthropogenic plant, tell-tale of soil formerly disturbed.

Colin went through reports of commissions into the state of the Forest and I looked for references to gorse in the literatures of farming and agricultural history. Nothing very specific turned up and, although in other districts gorse had been deliberately sown to heat bakers' ovens or left in small parcels by enclosure commissioners to supply the poor with fuel, references to it in the Forest were very few. The Abstract of Claims for 1853 recorded long lists of encroachments, including repeated infringements and continual destruction by the Forest keepers. Pressures had increased in the 1790s. These were the crimes of small men and the land they took in was small too, measured only in perches. Some of the enclosures we found were much larger. Documented infringements could not however be linked to sites on the ground and gorse went unmentioned. Three patches of gorse sown for rearing cattle and attracting good rents were however noted in 1856.

Given the almost total absence of documentation, I went to the Ordnance Survey, then at Chessington, to look at the files. On their maps every site we had located was pencilled in. The Ordnance Survey had missed nothing. At the end I stood and picked up my papers. 'What have you found?' asked the Librarian. 'Nothing, you people have beaten us to everything'. 'Oh no', he said, 'I just marked down everywhere you told us about in your letter'.

Another season passed before we found a means whereby the sites might be dated. I was able to persuade Dick Webster, a soil scientist and one of my fellow research students in the Department of Agriculture at Oxford, to take an interest in using soil pollen analysis to date early features. In turn, he persuaded Geoff Dimbleby of London University to join us in the New Forest and, as a favour, to collect cores of soil from a few of our sites for analysis in his laboratory. Because of the work involved in processing the material, and the fact that those on the science side were not primarily concerned with unravelling ad hoc historical puzzles like the settlement history of the New Forest, there was no chance of obtaining many dates. We had to make do with a handful. Naturally, or at any rate understandably, we selected places that seemed to represent a range of enclosure types.

To use soil pollen analysis for dating earthworks it is necessary to have a reference collection of plant species, the relative frequencies of which are known from other evidence to have changed over time with the climate and other factors. A limitation from our point of view was that there was no prospect of precise dating within the past four hundred years, which was the period of most interest to us and the span most likely to have tied in with any written evidence. Over that period the botany was muddled by introduced species. In the event the results emphasised the wide range of periods in which encroachments on the open Forest must have taken place, sometimes long before medieval monarchs reserved the area for hunting deer. Enclosures had been made over dates starting with the Iron Age and ending with the Middle Ages. The archaeology and land-use history were clearly complex.

Although it is not to be expected that the extension of cultivation on heathland outside the New Forest would proceed in the same way as within the Perambulation, the presence of essentially the same soil and vegetation, as well as the same price regime, suggested similarities. We had found two relevant early articles. They pertained to a project to take in so-called 'waste' in Christchurch parish, not far outside the Forest. Enclosure commissioners allotted 7000 acres there in 1803. Product prices were encouraging for heathland reclamation and we came across one other example, when T. H. Scott received a prize for taking in 170 acres at South Stoneham between 1808 and 1812.

The dates are significant because they mark the high tide of almost frenzied reclamation activity during the Napoleonic Wars. In Cornwall, Norman Hilton was convinced that this was a peak period for extending cultivation on the Land's End heaths, with some retraction afterwards. On the slopes of Dartmoor, Devonshire, a few examples of ridge-and-furrow from that period are still to be seen, because subsequent ploughing campaigns had never reached so high. How much of such marginal land was ever cultivated successfully is unknown. In *Small Talk at Wreyland*, referring to Dartmoor, Cecil Torr observes that 'enclosure is a mania that recurs at intervals; and deluded people think that, if they cut the moor up into fields, they will reap as much as in the valleys 1000 or 1500 feet below'.[2]

Along the Welsh border and in Wales there are also marks of ploughing; cereals were sown far up on Plynlimmon, although farms were abandoned there after 1815. Hills in the Wessex chalk lands similarly exhibit a little abandoned ploughing rig. And some of the derelict enclosures we located in the vicinity of the New Forest clearly stem from the same period. The enthusiasm did not last long. At Christchurch the local agricultural society discontinued its prize for heathland cultivation in 1812–1813 as no longer necessary. The landscape evidence in these fringe areas may be marginal but it is telling, coming as it does when documentary confirmation is so scanty.

The contemporary articles reported that many of the labourers who received allocations at Christchurch enclosed them during the following year, hiring others to throw up earthen banks. Gorse seed was sown on the sides and tops of the field banks to make hedges against marauding cattle or deer. Four of the new small proprietors built cottages for their workers and shelters for cattle far out on the heath. Some were in a great hurry to sow cereals without first planting fodder crops. They applied too little fertiliser, got poor results and gave up in disgust, bringing 'heath farming' into local disrepute. Because others among them were more successful, we can perhaps put the failures down to lack of capital rather than the inherent difficulty of cropping the heathland. The soil of the heaths varied; there were good parts and bad parts. One completely new farm was established where the heather was mingled with bracken and gorse, plants that were traditionally taken to indicate reasonable soil. Under heather lies lead, the adage went, under bracken lies silver and under gorse lies gold.

Measurements, diagrams and a map were supplied in one of the articles about Christchurch, which enabled us to locate some of the remaining banks. We were also able to find the banks between the long strips into which nearby West Parley Common had been divided. This had been occupied very much earlier than the 'waste' at Christchurch. The Lord of the Manor had been Sir Henry Weston, grandfather of Sir Richard Weston, the Surrey agricultural improver whom the ever-enthusiastic Arthur Young dubbed, 'a greater benefactor than Newton'. Sir Henry sold West Parley heath to two yeomen acting on behalf of all the tenants. But in 1619 the villagers complained to the Court of Chancery that two men were usurping the ground for turf-cutting and offered to crop it instead. Their proposal was successful, although the commissioners appointed to share out the land did not get round to reporting until 1633.

In 1929 a New Forest author, C. D. Drew, discovered an account of the division at Parley and mapped the layout from surviving boundary markers.[3] Two field names preserved those of the original allottees, which helped to confirm their positions. Long strips across a swampy valley had given each man a share of good and bad land for digging turf. Part of the southern end had been broken up and whereas, 'more have gone back to heath and furze and such apt names as "The Folly" and "Troublefield" tell their own tale', some good fields remained. What is evident outside the New Forest is that smallholder behaviour on the heath was characterised by opportune use. An extraneous event provides a sidelight on the fluctuating use of the land: in 1685, when the Duke of Monmouth, on the run after his abortive rebellion, was found huddled in a ditch on Horton Heath west of Ringwood, a description of the scene notes that while some intakes were in use others were grown over. This suggests an intermittent rather than permanent extension of cultivation.[4] Sustained cultivation might have been beyond the abilities of small men lacking enough livestock to produce the fertiliser needed to keep all the land in heart, quite apart from the fact that some land was taken in order to cut turf or peat for fuel.

The enclosures at Christchurch and Parley offered parallels to those on the New Forest heaths, the difference being the institutional one that within the Perambulation the Crown recurrently suppressed intakes.

The other pointers were all there outside the Forest: the use of bracken and gorse to suggest better soils than were likely beneath open heather moor; the dateable accounts of embanking; the sowing of gorse on enclosure banks; and the abandonment of some efforts at cultivation. Leaving aside for the moment questions about the actual use of the New Forest examples, the problem remained to date them. Very few commoners kept records and in their legally more exposed situation they were wise not to do so, because the Crown periodically held courts of inquiry to identify and throw down encroachments. Such written evidence as we did find stemmed from these attempts to throw out squatters, roll back their surreptitious advances, and stabilise the margin of settlement. Writers had long complained bitterly about squatters as idle, drunken, immoral poachers of deer. William Gilpin, clergyman at Boldre in the New Forest and propagandist for a new romantic view of the landscape, wrote in *Remarks on Forest Scenery* (1794) that squatters threw up houses overnight to establish a right to settlement. For a mile between Beaulieu manor and the Forest edge there was a settlement of cottagers, ostensibly cutters of gorse to fuel limekilns but in reality, Gilpin averred, poachers, deer-stealers and timber thieves.

Many of the enclosures we found were covered with gorse. From another perspective it might be said that the concentrations of gorse were bounded by enclosures. These were sometimes dateable by various, mostly oblique, means to the late eighteenth and early nineteenth centuries. Those out on the open heath seemed older. Enclosures were complicated in detail but in the former category typical examples lay alongside one or other of the straggling villages around the margins of the main bodies of heathland. Their banks bounded areas of, say, 500 feet by 300 feet, the banks themselves being perhaps seven feet wide, two or more feet above ground level, and four feet above the bottom of the outside, dry, ditches. Because the banks were smothered with gorse and the entire scene was low-lying, they were not at first sight easy to distinguish from the broader zones of gorse, which helps to explain why they had never attracted attention.

Inside the enclosures was gorse, mixed with grass or bracken. A few revealed narrow ridge-and-furrow, though not all bore explicit signs of cultivation. Soil disturbance had fostered a distinctive plant association. As a legume that can fix its own nitrogen, gorse is well adapted to colonising disturbed soil. We could tell by eye that at least three-quarters of the ground occupied by gorse had been subject to human disturbance in one form or another, including road works, old gravel workings and the scars of the military during the Second World War. To what extent the abandoned enclosures had once been planted with cereals and only taken over by gorse spreading from the banks after the little fields went out of use, as opposed to being sown with gorse for fuel or fodder in the first place, is a moot point: both, possibly, at different times.

The Crown's legal might—brought to bear intermittently but ultimately irresistibly—had frozen the fringe of enclosures, turning them into archaeology. The enclosures, derelict or not, do have an interesting historical message in that they hint at reserves of land cultivable with effort and capital. In short, it was not an absolute lack of land that held back agricultural improvement. Retreating cultivation, together with the paucity of sites right out on the open heath, suggests a shortage of inputs or lack of legal protection rather than inherent defects of soil fertility. The true potential was signalled by the success of the New Forest Pastoral Development Scheme, where 1000 acres were cultivated and re-seeded as grass leys between the years 1944 and 1948. Capital and labour were acutely scarce during that hungry time or even more land might have been taken into help alleviate food scarcity. The same was probably true of the Napoleonic and other early periods.

Heathland Ecology and Institutions

During the late 1960s I was living at Mortimer Common, Berkshire, and in the early 1970s still spending my summers there. Ten minutes' walk away was a group of Bronze Age barrows, probably indicating the first phase of clearing the forest and its transformation into heath. Close to the barrows were ploughing marks that seemed to be ridge-and-furrow left from an attempt at cultivation in the nineteenth century. The tracts of heathland hereabouts remained conspicuously undeveloped considering that they lie only forty miles or so from London. What would be expected, theoretically speaking, is to find London surrounded by rings of land, each less intensively cultivated the further it was from the city. These bands are known as von Thunen rings and imply that transport costs ensured that land closer in, being more accessible, was more valuable and more intensively used than land further out.

In reality the concentric pattern breaks up. Tertiary heaths come close into the city: Cobbett complained of the 'villainous heaths' and Hounslow Heath was a notorious resort of highwaymen. Heaths can poke through modern exurbia. These stubby fingers of land almost always seemed too low-yielding to repay permanent cultivation yet surprisingly often have escaped the builder as well. The heaths along the Berkshire-Hampshire and Surrey-Hampshire borders are of this type, retained in the past mainly for grazing and afterwards for growing timber. Here ecology trumps the power of county boundaries. 'In the real world', so the garden designer, Gertrude Jekyll, put it, 'one steps without knowing it from Surrey into Hampshire on the dry heath-land, and out of Surrey into Sussex from one clay puddle into the next, without ever being aware in either case that the land is called by another name'.[5]

Counties are ecological ragbags, gathering within age-old historical units motley ecosystems with nothing natural in common. County institutions were once influential, such as the 'custom of the county' which specified which children would inherit a farm. But political categories at regional or level levels were not everything, ecological categories could be influential too. As John Hutchins observed, Jekyll-like, in 1774, 'there is another [type of division] not yet described; I mean that which nature has formed. The others are arbitrary and unstable; this is fixed and invariable. Whatever changes a country undergoes, the soil and situation are still the same: and here the down, the vale and the heath are always distinct: nature has set to each its proper bounds, which it cannot pass'.[6] Hutchins was the quintessential Dorset author. Another old authority on Dorset wrote, 'we trod the rich soil of the vale of Blackmoor until we came to Revel's Hill, before we searched into the nature of the soil we were sufficiently *instincted* that it had passed into the chalk by the altered aspect of the vegetation'.[7]

In Cornwall, Norman Hilton suggested looking at moorland fringes to determine the possibilities of reclamation, basing his conclusions on field surveys and distinguishing categories of use and re-use on a map entitled, 'Changes in the Margin of Cultivation c.1485–1952'.[8] The pertinent point is that, unlike Jekyll or Hutchins, Hilton discounted soil differences or inherent fertility in explaining the curious mixture of farmed and abandoned land. He claimed that the distribution of cultivated land was dominated by historical contingency, perhaps related to the early distribution of settlement rather than ecological dictates.

Hence ecology and institutions do not invariably produce coincident distributions. Gertrude Jekyll's county boundaries have no obvious relationship to land use, but the Perambulation (legal boundary) of the New Forest can make all the difference in the world. At Holbury Purlieu the

Perambulation cuts sharply across the heath, leaving unmodified heather moor on the Crown land of the Forest but permitting the cultivation of private land on the other side. It is a stark, artificial, institutional outcome, which reflects ancient historical decisions. And ecology also took second place in Cornwall. Farmers and gardeners are ordinarily well aware of soil differences, which under any given set of techniques translate into relative levels of cost and size of yield. Into this institutional factors (such as the Perambulation) sometimes intrude but they do not do so everywhere (county boundaries), while in Hilton's Land's End private initiative overrode all else.

The adaptation of farmers and cottagers to their circumstances bears examination. The matter needs to be taken case by case, requiring sensitivity to ecological factors important to people working with their hands in the laborious circumstances of the past, but which today may no longer be significant or even recognised. Consider an occasion when Edward Lisle, who lived at Crux Easton, Hampshire, over the turn of the seventeenth and eighteenth centuries, criticised some workmen for threshing too little grain in a day.[9] They persuaded him that the task varied from place to place, because different soils produced different lengths of straw. The long straw of the hills meant that fewer sheaves could be handled on the threshing floor at any one time than straw from the valleys; only half the quantity of wheat could be threshed on the hills in a day. This wide gap, rooted in the natural world, has been steam-rollered into insignificance by modern machinery and agricultural chemicals.

People in the past were acutely conscious of ecological reality, which does not mean they wrote down things that they took for granted. Light and heavy land favoured different products and were not equally easy to work. One farmer's wife in the light Suffolk Sandlings was spoken of admiringly as 'a fine cheese-maker – a *heavy-land* woman she wor'.[10] Lisle, too, engaged in a discussion about making cheese, in particular querying why Somerset cheese was not as good as that from Northwest Wiltshire.[11] He was the complete economist who understood the substitution of factors of production, arguing that if the difference in quality resulted from the different skill of the makers, Somerset men would have imported wives and dairymaids from Wiltshire!

The three main divisions of southern England, light land, clayland and heathland, are highly distinguishable. (A fourth division, hay meadow, which is a subset of the clay vales, is also worth notice as Chapter 9 will show.) The natural divisions influenced the costs of producing different crops or livestock, besides bearing on the perquisites of the countryman, notably the vital question of how much wood he had for fuel. Geological outliers and inliers, gradations of soil type and topography, all introduced complications. Soil texture can be amended by marling, putting clay on light chalk soils or lime on the heavy clays. These additives had to be continually replenished—a labour-intensive business. The soil was being modified rather than made completely uniform, which was beyond the power of the past.

Furthermore parishes and farms can overlap more than one soil or angle of slope and be laid out (as parishes had been in Saxon times) in order to give individual units a portfolio of options. The arrangement provided insurance against the hazards that might reduce production from a single type of land especially hit by adverse weather. In other words, significant as the underlying ecology was (more significant than is usually realised today), institutional distributions could deliberately seek to offset the harm that might accompany reliance on a single type of land. Direct costs of production are not however the only considerations. Geographical groupings affect outcomes by influencing transaction costs, as has been shown by studies of differences between grids and irregular divisions in the American landscape.[12] Man-made divisions can thus override ecology; institutions are 'sticky' and determine the distributions of economic activity where the decision has been made to do so. Where historical contingency intervenes there is no unique rule.

Heathland Exploitation

The heaths were distinctive. Soils on the Tertiary gravels were even lighter than those on the chalk and limestone. In some compensation, they grew wood that provided material for craft industries, and contained gravel for the digging, corners for fishponds, sites for beehives, turves for firing, gorse for fuel and fodder, besides grazing and browse for geese, cattle and ponies. An accumulation of these minor resources meant that 'peasants' were able to make a living of sorts on heathland held in common.

The value of the heaths cannot be judged by the uncultivated remnants today. Much has been enclosed. Yet the soils were not good enough to be a great draw-card for manorial lords and where there were enough small men they could sometimes band together to defend their access to the bundles of common rights on which their humble livelihoods depended. The survival of cottage stockkeepers living off local resources—hundreds survive in the New Forest, which makes it special—thus has an institutional as well as environmental origin. In the nineteenth century, the commoners of the Forest organised against the Crown's intended afforestation of grazing land. They had, and retain, the advantage of living within the Forest Perambulation, where an administration able to control the use of the heathlands on the open Forest exists, with the legal power to exclude encroachments by other commoners, incomers and would-be improvers alike.

Where small, independent men were scattered and fewer in number, as along the Berkshire-Hampshire border, they have almost vanished. They had no central administration to be turned in their favour. And once their heaths were no longer grazed, the land went back to scrub and grew into woodland, notably silver birch. Previously the commoners cut the gorse and burned or swaled the heaths in late winter to encourage the growth of *Molinia* grass, which is palatable to livestock. The lack of regular burning is why the vegetation becomes coarse. From the naturalist's point of view former commons all across the South desperately need burning, but a combination of local authorities more concerned with public relations than the environment (regardless of what they say), a constituency of residents who have been permitted to build closely around the heaths, and aggressive dog-walkers, impedes sensitive management.[13]

In the Hampshire-Berkshire borderlands there were still a few old besom (brush) makers at the end of the 1960s and I visited one rake-maker in his ancestral cottage. Now only the name 'Rakemakers' bestowed on his house remains to remember him by. Woodland crafts are also faintly preserved in the name of a public house at Mortimer Common, the 'Carpenters and Turners Arms'. Charcoal burning continued nearby into the 1970s. But by that date all connection with heathland farming was lost.

By contrast, over one thousand holdings, mostly under 50 acres, some hundreds of them owner-occupied, still represent the main source of livelihood to their owners in the New Forest. This community is of long standing. In 1875 G. E. Briscoe-Eyre quoted evidence to a Select Committee of the House of Commons demonstrating that common rights were essential to the cottage stock-keepers. He described them as a better-off labouring class who seldom rose or fell in position, though he mildly contradicted this by mentioning a few who had clawed their way up the social ladder to the status of tenant farmer.[14]

Briscoe-Eyre was frustrated, as all students of the New Forest commoners have been, by the difficulty of finding out about their financial situation—not that there have been many students, because the commoners barely figure in the national economy.[15] 'The cottagers are exceedingly reserved, and information is difficult to obtain', wrote Briscoe-Eyre, 'besides, few of them keep accounts…' Their earnings from jobs away from the holdings and incomes from common rights were hard to separate. He made an effort to calculate the profits of stock-keeping by isolating money transactions and deducting any costs he knew about. He also recorded the profits of twelve month's stock-keeping on a 'little place' where an exceptional man did keep accounts.

Briscoe-Eyre's opinion was that, 'the cottager lives the life of a labourer, but the profits of his little holding will compare with those of a farm in an enclosed country of about thrice the size, and of about twice the rent' (p. 51). Bankruptcy and borrowing from money-lenders were unknown. The aged were well enough off to keep out of the workhouse and there were cottagers who were more prosperous than skilled, fully employed, London artisans. Rights of pasture on the open Forest were what supported the commoners. Other rights and perquisites were allowed by the Crown and 'nothing is wasted, and even the furze-tops are gathered in winter, and cut up with the other food' for the cattle (p. 47). By and large Briscoe-Eyre's conclusion still rings true. The survival of such people is remarkable given that the New Forest is squeezed between Southampton and the urban sprawl of south Hampshire on one side, and Bournemouth, now in Dorset, on the other.

London and the East Hampshire basin are prominent among the locations of heathland and have been studied as closely as anywhere.[16] This may have been the area of Cobbett's villainous heaths and described by fearful travellers and agricultural improvers alike as the 'waste' but the reality was that productive use had always been made of the land; many products came forth and fights over grazing rights could break out between villages. Howkins in *Gorse, Broom and Heathland* consciously ignores the most minor examples but still offers a long list of uses. They may be grouped under the common rights that facilitated their production: estovers gave the right to cut heather and other plants; pannage the right to run pigs; and common in the soil was the right over inanimate resources, mostly for gravel digging. The gorse that grew on Surrey heathland might even be accepted in lieu of tithes, because it was saleable as fuel and fodder. During the seventeenth century it was cropped and gorse seed was sold in London. Bracken was used for animal bedding, heather to roof the poorer dwellings, and gorse for firing and animal feed.

Nor was the use of heathland products merely a matter of 'peasant' self-sufficiency, although it was that too, because there were markets for gorse. Different parts of the plant had different uses and were priced separately. All told the gorse economy was far from simple or trivial.[17] Because the modern world has no comparable use for the species—and agricultural historians mainly restrict themselves to the precursors of current crops—the earlier significance of gorse has escaped scrutiny. Yet as a fodder gorse has half the protein value of oats, which meant that commoners did not overlook it.

Stepping a little further west, Oxford bakers and colleges imported gorse from Eynsham heath by river, down the Thames. There and elsewhere urban demand was mainly for fuel, particularly in bakers' ovens, with the concomitant risk of fire when something so flammable was stacked inside villages and towns. The Great Fire of London started at a baker's in Pudding Lane and many fires elsewhere had a similar origin. As a further aside, Eynsham Park was laid out on the heath after enclosure, a frequent enough occurrence when commoners were unable to protect their rights.

On the Surrey heaths little intakes were surrounded by earth walls surmounted by gorse hedges like the enclosures on the New Forest, where we have to suppose they were similarly constructed to grow gorse for fuel and fodder. But whereas the Forest was regulated by the Crown, and encroachments continually thrown down, Surrey was not legally protected. There, large-scale enclosure went a long way towards disrupting the rural economy; afterwards pine plantations were grown, blanketing the heath and excluding previous uses. A few new farms were established and the heaths colonised by these 'extension' (of cultivation) farms in a process reminiscent of reclamation on the chalk downs and Cotswolds in the eighteenth and nineteenth centuries.

Today only fragments of the formerly extensive heathland remain. Land grabs to privatise them by manipulating the laws regarding commons have sometimes succeeded and remain threatening.[18] Everywhere, the characteristic birds and plants have diminished in numbers, insects more so. Well-meaning efforts to conserve the heaths by employing grazing animals to suppress the scrub appear to be too heavy-handed for the recovery of a diverse insect fauna. Even in the New Forest grazing pressure is considered to be excessive, although because it is now a National Park it is better protected in other respects. Elsewhere, the cessation in the mid-twentieth century of the regular burning and calibrated grazing which were features of past management has often permitted birch to take over, converting heathland into woods with paths interesting only to dog-walkers. Details of the historical economy are limited and fragmented but a composite picture may still be drawn. It shows that lowland heath was not a timeless ecosystem but an artefact of long-standing uses which have now been superseded.

Exploitation at East Woodhay

It is sometimes said that humanity once lived surrounded by nature but has extended its grasp so greedily that it is now nature that is enclosed. Into the nineteenth century there were however some remaining 'islands' of less developed land which tempted the improvers of the period into energetic assault, Otmoor and Wychwood Forest being prominent examples. A lesser known case lay in the parish of East Woodhay which stretched from the southwest of Newbury almost to the Wiltshire border. East Woodhay Common was originally a tract of 1200 acres on which a few young cattle and a few poor quality horses were fed. Improvers of the 1790s decreed it of little value, which meant they discounted its value to the local cottagers. Today it is an area of prosperous dwellings, each with generous grounds, set tastefully amidst fields and woods. An investigation of its history is needed to perceive that this currently exurban territory was an untamed 'island' of heath before it was masked by agricultural and residential development.

East Woodhay consisted of secretive little clay bottoms, heaths on intervening tops of Tertiary gravel and a fair covering of woodland, lying a mere half-dozen miles from Newbury. Access was difficult, especially in a wet winter. That somewhere so close to a town and to the Bath Road was thought-out-of-the-way seems almost ludicrous, yet the parish was of little interest to polite society until its potential was spotted by an entrepreneurial (and 'patriotic') clergyman, Rev. John Harvey Ashworth. From 1830 into the 1840s he bought a series of old farms in the vicinity, developed them into small estates, improved the houses or built new ones, and sold them off.[19] They were not the centres of landed estates but bourgeois mansions with just enough land to guarantee privacy.

The medieval church at East Woodhay had already been replaced in the 1820s by a large red brick edifice, something out of the ordinary in a district of small medieval stone churches and an added attraction to prospective residents. Ashworth subsequently created a series of seven or eight smart villas, each in its own grounds. That he was able to find 'proprietors of fortune' to buy them shows how much non-agricultural wealth there was by the second quarter of the nineteenth century and how eagerly it was seeking homes in the countryside. The residences are now hard to see behind their earthen banks, walls, screening trees and holly bushes. Curving driveways conceal them and afford the passer-by barely a glimpse. The common had not merely been, as it were, stolen from the goose but hidden away to boot. The area now gives the impression of a slice cut out of extra-urban Surrey and cast down in West Berkshire.

Ashworth lived in one of the bigger houses. Despite the fact that he was praised as 'distinguished alike by his high literary acquirements and the unceasing efforts to promote the welfare of all who come within the sphere of his influence', a more candid age might think of him as a land speculator. He was temperamentally a booster, writing a book about Ireland at the height of the Famine extolling its developmental prospects! His benevolence may be judged from the letter he wrote to the Earl of Carnarvon, whose chaplain he was, objecting to the use of a parcel of Woodhay land as allotments for the poor.[20]

Modification of the local landscape continued. A surgeon, John Winterbottom, bought Tile Barn Cottage and after he had married a healthy widow (sic) converted it into Tile Barn House.[21] He had the road through Woolton Hill diverted so as not to pass his door; the old road became his driveway. This fetish for privacy among the prosperous inspired similar 'road capture' right across the country.

Local people had been lawless, or from another point of view independent, but into their home district the new resident proprietors spread, 'happiness and contentment around them... benefiting in no small degree, the commerce of the neighbouring town of Newbury'.[22] The construction set afoot by Ashworth and those he enticed to buy into his scheme required immediate but perhaps not lasting labour for pursuits such as burning brick and tile; the distinctive local Hollington brickwork occurs in several houses hereabouts. Other men could find employment as gardeners and outdoor staff and the women as domestic servants, all very different from their forebears and far more consistent with the spirit of Victorian social order. They were tamed as their heaths and woods were tamed.

Modern Reclamation

By the late twentieth century surviving heaths outside the New Forest were degenerating badly. One person to resist was the outstanding West Berkshire naturalist, L. R. Lewis.[23] The ignorance of local bureaucrats and the way they sanitised their procedures with a mantra of 'the environment' infuriated him. He was scandalised by the way the tertiary heathlands were being permitted to turn into birch scrub. This outcome was on the cards because the old grazing regimes and the periodic burnings by which commoners had encouraged an early bite of *Molinia* grass had ceased. Without regular burning, the open spaces were certain to grow into woods, blocking out the Nightjar and other heathland wildlife species. The authorities thought the public would be content with woodland paths, along which they could trail their dogs. Politically they may have been right, but historically and ecologically, not. They were banking on acceptance of the 'Crete Scenario' where the public will tolerate a barren environment around them provided they can watch nature films on the television and take their children to zoos for a sight of exotic animals.

Lew Lewis's anger at the mismanagement of the heaths came to a head when the U.S. Airforce quit the airbase on Greenham Common, Newbury. The runways were bulldozed into mini-mountains of concrete fragments, trapping heat in summer. His first thought was that this was an outstanding chance to experiment: let's see what a really warm habitat can produce in Berkshire. That was too creative for the bureaucrats, who ordered the heaps to be cleared and the public let in, adding insult to injury by commissioning congratulatory messages about 'preserving the environment'. The public brought its dogs and drove off the ground-nesting birds.

Greenham Common became what Lewis labelled a 'pooch park'. His observation that it had been ecologically more rewarding when occupied by the United States Air Force was not popular. For him, the last straw was when a television 'celebrity' was drafted into declare what a splendid nature reserve had been made of Greenham. He fired off a salvo in the *Newbury Weekly* News, but as is the way with land management in England the authorities took no notice. In November 2009, only months after his death, an advertisement appeared for three part-time wardens to protect ground-nesting birds at Greenham. This splurge of taxpayers' money would not have been needed had the area remained forbidden to dogs from the outset. But in the Parkinsonian world of modern environmental bureaucracy a single old naturalist with decades of faithful recording behind him had no *locus*, as the lawyers say. Heathland, once a fully functioning economy and ecosystem, has been reduced to shreds, inexpertly managed by precisely the authorities who are charged with its management.

Notes

1. James Willis, 'On Waste Land', *Communications to the Board of Agriculture* VI, Pt. 1 (1808), pp. 16–30, and Willis, 'Communications on Fences' VI, Pt. 1, pp. 237–251. For the New Forest proper, see E. L. Jones and C. R. Tubbs, 'Vegetation of Sites of Previous Cultivation in the New Forest', *Nature* 198 (8 June 1963), pp. 977–978, and C. R. Tubbs and E. L. Jones, 'The Distribution of Gorse (*Ulex europaeus* L.) in the New Forest in Relation to previous Land Use', *Proceedings* of the Hampshire Field Club and Archaeological Society, XXIII, Pt. 1 (1964), pp. 1–10. This work was developed with respect to soil pollen analysis in C. R. Tubbs and G. W. Dimbleby, 'Early Agriculture in the New Forest', *Advancement of Science* (June 1965), pp. 88–97, with respect to Dartmoor by John Dearing, 'Gorse, Man and Landuse Change on Dartmoor', *Transactions* of the Devon Association for the Advancement of Science 109 (n.d.), pp. 135–152, and with respect to soils by K. S. Reynolds and J. A. Catt, 'Soils and Vegetation History of Abandoned Enclosures in the New Forest, Hampshire, England', *Journal of Archaeological Science* 14 (1987), pp. 507–527.
2. Cccil Torr, *Small Talk at Wreyland* (Oxford: Oxford University Press, 1979), II, p. 31. He makes further comment about the repetition of such schemes.
3. C. D. Drew, 'The Division of the Commons of West Parley and West Moors', *Proceedings of the Dorset Natural History and Archaeological Society* 50 (1929), pp. 109–116.
4. Alan Ivimey, *Pilgrim's Pleasure: The West Country* (London: Newnes, 1959), p. 154.
5. Gertrude Jekyll, *Old West Surrey* (London: Longman, Green, 1904), p. vii.
6. John Hutchins, *The History and Antiquities of the County of Dorset* (1774), p. lxx.
7. Quoted by L. H. Ruegg, 'Farming of Dorsetshire', *Journal of the Royal Agricultural Society of England* 15 (1854), p. 399.
8. Norman Hilton, 'The Land's End Peninsula: The Influence of History on Agriculture', *Geographical Journal* CXIX (1953), pp. 57–72.

9. Edward Lisle, *Observations in Husbandry* (London: Hitch et al., second edition, 1757), I. p. 345.
10. George Ewart Evans, *Ask the Fellows Who Cut the Hay* (London: Faber & Faber, 1965 edition), p. 69.
11. Lisle, *Observations,* pp. 146–147.
12. Gary D. Libecap and Dean Lueck, 'The Demarcation of Land and the Role of Coordinating Property Institutions', *Journal of Political Economy* 119/3 (2011), pp. 426–467; Deirdre Mask, *The Address Book* (London: Profile Books, 2020), p. 113, citing a Harvard Ph.D. thesis.
13. In *Newbury Weekly News* 27 Apr 2006, L. R. Lewis exposed the weak management of Greenham Common, Berkshire, and the incongruous self-praise of its 'restorers'. He pointed out that those responsible for the Common had not even established a baseline from which to measure wildlife changes.
14. G. E. Briscoe-Eyre, *The New Forest: Its Common Rights and Cottage Stock-keepers* (Lyndhurst: J. C. Short, 1883).
15. C. R. Tubbs, 'The Development of the Smallholding and Cottage Stock-Keeping Economy of the New Forest', *Agricultural History Review* XIII (1965), pp. 23–39, is the classic modern paper by someone who knew many commoners personally.
16. Useful sources for heathland history west of London and in East Anglia are Mary Ann Bennett, *Life and Work on Surrey Heath* (Chichester: Phillimore, 2007); Lee Chadwick, *In Search of Heathland* (London: Dennis Dobson, 1982); and Chris Howkins, *Gorse, Broom and Heathland* (Addleston, Surrey: Privately printed, 2007).
17. The use of gorse was on a sufficient scale for a 'gorse or furze-bruiser', powered either by hand or by horse or water, to be illustrated in Mary Wedlake's *Price List of Modern Farm Implements* (London: n.d. [1850]). The list was sent out with that year's volume of the *Journal of the Royal Agricultural Society of England*. An accompanying sales pitch rhapsodised about gorse as an animal feed. Howkins, *Gorse, Broom and Heathlands,* pp. 33ff. describes the harvesting, transportation and sale of gorse with emphasis on the skill required to deal with such a prickly and bulky commodity.
18. *The Countryman* (95/3 Aug–Sept 1990): editorial.
19. Bernard Burke, *Family Romance; or, Episodes in the Domestic Annals of the Aristocracy* (London: Hurst & Blackett, second edition 1854), I, pp. 138–139; *Letters & Papers of the Bath & West Society* VIII (1796), p. 148; and East Woodhay Local History Society, *A History of the Parish of East Woodhay* (Privately printed, 2000).
20. Carnarvon of Highclere Papers, Hampshire Record Office 75M91/25, c.1833.

21. E Woodhay Local History Society 2000; Orton House, Woolton Hill, Draft Design and Heritage Impact Statement.
22. Burke, *Family Romance*, p. 139.
23. L. R. Lewis, *An Introduction to Landscape in West Berkshire* (Newbury: Privately printed, 1998).

CHAPTER 7

Lot Meads

Yarnton Mead

The formerly widespread system of allocating rights over wet grassland by the drawing of lots has rather seldom been studied. Even the lot-drawing ceremony at Yarnton, four miles from Oxford, which was still operating in the 1960s, was ignored by the numerous historians in the city. My own interest was piqued when I heard about it from local people and for several years during that decade I went to the ceremony and took photographs. Since then a few local historians in various regions have occasionally paid attention to lot meads. Typically their research has been carried out as part of the history of river meadows as a whole and has tended to concentrate on botanical issues as offshoots of the rising concern with conservation in recent years. Aspects that an economic historian might think pertinent get short shrift and questions about broad historical matters such as the inception of the system, its productivity and its demise are seldom directly asked. A weakness is the lack of comparative studies: single cases tend to monopolise attention. Nor have scholars made much attempt to capitalise on the last handful of cases where communal methods of allocation survive, or did until recently. Before discussing general issues I shall therefore consider three specific instances.

The first example is the lot-drawing ceremony at Yarnton Mead itself. This was the last of the Thames meadows to function as a lot mead.

E. L. Jones, *Landscape History and Rural Society in Southern England*, Palgrave Studies in Economic History,
https://doi.org/10.1007/978-3-030-68616-1_7

When R. H. Gretton wrote about it in the *Economic Journal* for 1910 it was already a rare survivor (he thought it unique).[1] The ceremony and related management practices have vanished now. On one occasion in its late years, in the 1960s, my daughter, then a very small child and obviously independent of the participants, was asked to make the draw. She is probably the youngest person still living to have played this role in a procedure that may have gone back as far as Saxon times. Because the ceremony was ancient and continued in recognisable, if attenuated, form to the bitter end; because the mode of management had special ecological as well as historical interest; and because no other historian was ever present, it seems worth recording my experiences.

Yarnton Mead is a mainly level site of 80 acres of grassland, but with subtle variations in topography and vegetation. Some of the strips into which the mead was divided had better soil than others; some were prone to flooding but might do better in dry years. People who have made hay by hand confirm how growth varies between seasons and how significant small differences are. The variations have a real impact on the work load, for instance, determining the number of times the hay has to be turned. To explain the systems of strips and lot-drawing, agricultural historians freely attribute to our ancestors the motive of averaging out the differences, offering each person insurance against too wet or too dry a year. From what those carrying out the ceremony told me on the spot, this seems to be true. It all sounds nicely egalitarian as well as pragmatic and presumably worked in a rough and ready way among those who actually owned the rights, although long before the 1960s most people in Yarnton village had ceased to own any rights at all.

The Mead was never known to have been ploughed, deliberately fertilised, or re-seeded, though the paucity of records means we cannot be sure of this. Nevertheless the Mead had not quite escaped botanical modification which potentially had agricultural significance. First, it had long been mown and grazed, conceivably for more than a thousand years. One study showed that the plant species differed from those of Port Meadow—the purely grazing meadow to which the Victorian medievalist, E. A. Freeman, used to take visitors when they asked to see the oldest monument in Oxford.[2] Secondly, introduced species such as clovers and greater salad burnet (another forage plant) had invaded from adjoining land. Thirdly, a dilute solution of fertiliser, and perhaps more significantly loads of silt, had washed from upstream fields during winter floods. Fourthly, flood control higher up the Thames had altered the water regime.

The Mead contained more herbs than other Thames-side meadows but a smaller percentage of nutritious grasses. The yield was better than quality. This was reflected in the low price of the hay, which would have soon caused milking cows to dry off. One Meadsman told me it was deficient in manganese and he always sold his Yarnton hay. Store and dry cattle loved it but it did not make them grow much and had formerly been sold to horse keepers. Until the 1920s almost every shop in Oxford kept a horse and cart and the hay fetched at least £4 per acre, although during the depression this sank to 1s.6d per acre. In the 1960s the price was £2.50s, held this high only because Oxford hospitals were willing to buy it for their experimental rabbits. In 1962 a Polish smallholder paid £4 per acre because he was short of feed but that was an exceptionally dry year.

The floristic variety, including a few scarce species, had led to designation by the then Nature Conservancy as a SSSI (Site of Special Scientific Interest). Unfortunately this meant only that the Conservancy would wring its hands at the Mead's destruction. When the late Professor David Chambers suggested the Ministry of Agriculture might like to help preserve the ceremony, as it had preserved the one surviving open-field arable village at Laxton, Nottinghamshire, the scornful response was that the Mead was in no danger.

The lotting procedure was as follows: six-acre lots, or fractions of lots, in Yarnton and nearby Pixey Meadow were auctioned at 'The Grapes', Yarnton, on the first Monday in July. Pixey's hay and grazing were said to be better than Yarnton's. What the bidder was buying was not a fixed piece of ground but one season's right (called a lot) to cut hay from a named strip of grass in the Mead. My marked catalogue (which is only a cyclostyled sheet) for 1 July 1963, lists lots or part-lots including three-quarters of William of Bladon, half of Watery Molly, and one-quarter of Rothe. Gretton traced eleven of the names in Yarnton records of the period 1066–1277.[3] Also auctioned were fifty commons or rights of grazing after the hay had been carried: one common permitted one cow or half a horse to be grazed on the after grass from 31st August to 1st December, following which all stock had to be removed because in wet winter weather they would poach the ground, that is trample it into mud, and harm the next season's growth.

A drawing of lots took place out in the Mead one week after the auction. Small balls made of cherry wood, each with a lot name painted on it, were brought in a canvas bag and drawn on the spot at each of the five divisions of thirteen strips. The individual whose ball was drawn first was entitled to the hay from the first strip in the division. He scythed a patch in the grass and cut his initial in the turf. The edges of the strip were then walked in single file from one corner post to the next. This was called 'running a tread' and was done to mark the edges of the strip. It was mildly surprising because inevitably some grass was trampled down.

The first strip of the first division having been drawn, the ball was not replaced in the bag, but a second drawing was made. A new ball was picked out, and so on until every strip in the division had been allotted to one or other bidder. All the balls were then put back in the bag and drawing started again at the next division, with the scything, cutting of initials and treading of the edges repeated at each strip. The result was that every lot holder ended up with a strip in each division—more if he had bid successfully for more than one. Where an individual had bid for a fraction of a lot, the subdivision of the strip was left to the two Meadsmen. The strips were subdivided lengthwise, to let the scythesmen work up a rhythm to their swing. It is interesting that at Aston and Cote, not far away, strips in the common meadow were often inconveniently dispersed and too narrow to scythe efficiently.[4] Tenants there often exchanged the lots they had drawn, which testifies to a measure of flexibility in a system where the agreed governance was normally rule-bound. Without the detailed contemporary report available about how things really worked at Aston and Cote such private arrangements would remain unknown; they can be only guessed at in most places, leaving the scholar a prisoner, so to speak, of any purely formal descriptions that survive.

At Yarnton, after the drawing, a date was fixed for starting to mow. The hay had to be cut and carried within about six weeks after that. This is where the real difficulty started. Just as the micro-topography of meadows varies, so growing seasons vary from one year to the next. Whereas a private land owner may cut earlier or later than his neighbours, according to his judgement about the hay's ripeness and the upcoming weather, there is no such latitude in communal farming. A single date is set for all. In 1962 the Mead was cut later than nearby fields, which had been mown quickly lest the dry season not last. The Common Mead's hay had to wait for the pre-assigned date, hoping that the weather would remain favourable.

During its last years the ceremony continued solely because a handful of individuals took a special interest in it. Chief among these was Mr. Edward Harris of Paternoster Farm, Yarnton, the head Meadsman, who had been involved without a break since 1921. In 1962 the only people who turned up, apart from me, were Ed Harrris, the other Meadsman who was called De La Mare, another retired farmer, and an old farm-hand who, in the hope of beer money, scythed expertly and ran the tread, despite having had his calves shot away during the First World War.

By the 1960s the continuation of the system had come to depend on Ed Harris's interest and knowledge. Even he found it hard to locate the corner posts of the strips in the tall grass. Demand for the hay had fallen away and interest in the ceremony too. In 1963 ten and one-quarter of the thirteen divisions were auctioned off to whoever would bid. In 1964 the field would not have been cleared at all had not Harris and one other farmer stepped in and bought most of the hay. In 1968 all thirteen divisions were put up for auction, there was only one bid, and the hay could not be cleared. As far as is known, that was the first year since its remote beginning when there was no lot-drawing ceremony. I had not been able to attend the auction but turned up in the Mead on the appropriate morning a week later. No one was there. I went to Paternoster Farm and Ed Harris told me that the Meadsmen could hope only to let the Mead for grazing. One week later a flood put paid even to that.

The ceremony did continue for a few more years but on Harris's death in 1985 none of the public bodies that might have been expected to do so came forward to help preserve the form of management. Two new Meadsmen took office but by the 1990s no one was showing the least interest in the hay. The Mead had to be let solely for grazing, which barely covered the cost of fencing and could not be depended on to preserve the precise flora.

Purton Stoke

The next example relates to the Poor's Platt charity at Purton Stoke, north Wiltshire. Charles I had granted the 'second poor' of this hamlet a piece of land measuring twenty-nine acres in compensation for the disafforestation of Braydon Forest, where they had possessed ancient rights of grazing cattle and gathering wood for fencing and firing. The second poor were people just able to support themselves without parish relief. Cromwell confirmed the grant in 1657 but soon afterwards the overseers of the poor of Purton, in which parish Purton Stoke lies, appropriated the rights and applied the revenue to reducing the parish rates.

In the 1730s the poor of Purton Stoke eventually asserted themselves, although their reasons for waiting so long are unknown. They brought an action to recover their rights and hired an expensive London lawyer to plead their case. This was heard in the upstairs room of The Bell Inn by no less a personage than the Attorney General. The depositions of the witnesses before him are on record. The first witness seems to have been a process server who came from Middlesex and had already confronted the occupant of the land with copies of the original decrees in the Court of Exchequer. Old inhabitants testified they knew of the grant, and that, although the poor of Purton Stoke may have derived a little incidental benefit, they never had the whole of it. One added that the plot had been enclosed with a quickset hedge at some unknown date. The actual occupier protested that he was on a yearly tenancy from the minister, churchwardens and overseers of the parish and did not believe that only the poor owned title to the proceeds.

Judgement was in favour of the complainants. No doubt the Attorney General was swayed by the existence of formal documents, which would make a stronger case in the eyes of a lawyer than the oral testimony of old men, though in most parishes their word alone ('Shepherd's memory') was all there was by which to ascertain boundaries and rights of way. The outcome made Purton Stoke one of a small minority of cases where poor commoners took the costly step of banding together to secure their rights, and one of the even smaller number where they prevailed. The Purton Stoke charity continues to the present day and the trustees are the custodians of minute books in which the first entries date from 1735, with details about the upkeep and letting of the land, and payouts to the poor listed from that year to the present.[5]

On rare occasions other cases did succeed and there were exceptional parishes where the poor were hot for their rights. One was Potterne, also in Wiltshire, which was described by an Assistant Commissioner of the Poor, as 'filled with a very discontented and turbulent race'. These men were ironically termed the 'Potterne lambs', who 'excelled in work, drinking and fighting'. The paupers of Potterne even raised a subscription among themselves to buy a copy of the law book called *Burn's Justice* 'for the avowed purpose of puzzling the overseers and magistrates'. Potterne was a parish with many hands from its former clothing industry, whereas an agricultural hamlet like Purton Stoke was scarcely big enough to generate so many barrack-room lawyers or such shop-floor camaraderie. Local willingness to defend the rights in the 1730s is the more surprising.

Until the Second World War, the trustees of the Purton Stoke charity met to receive bids for the ensuing year's grazing in the upstairs room of The Bell Inn, where the Attorney General had presided. The farmers passed around a bellows—presumably because it had been handy on the first occasion and became customary—on which each chalked his bid. The highest bid secured the land. Now the charity merely lets the land for between £1000 and £2000 per year and distributes the proceeds to those residents who care to register as worthy of receiving them. To distribute the payouts the trustees meet in the pub on the first Thursday after Old Christmas Day (6th January), as they have since 1745.

For several years during the first two decades of the twenty-first century, I was able to sit on the sidelines and watch the distribution. On a typical occasion the nine trustees dined at 6 p.m. and chattered away until 8 p.m., when they rearranged themselves around the table and the chairman made a little speech introducing the charity. He asked registered recipients to come forward (and re-register afterwards). By this time a few men had slipped in and were standing quietly to one side. The chairman called out a dozen names but only five came forward and took the cash. It was over in ten minutes. The others who were registered were apparently elderly and had not come; they may have received their money later, although 9 p.m. was the official cut-off time for making a claim. Any money not claimed goes back into the pool for future distribution. Throughout the ceremony a dozen younger inhabitants were propping up the bar at the other end of the pub's single room, taking no notice of what was going on. Although the edge of Swindon, a very large town aspiring to city status, is a mere four miles away no outsiders were present other than my wife and me and in one or two years our daughter.

Luppit Common

The third case is a more active microcosm of communal farming. It concerns the yearly meeting of the trustees of Luppit Common on the Blackdown Hills in Devon. The antiquity of Luppit village as well as the continuity of its life was plain from the fact that in the late 1980s, when I attended a meeting of the trustees of the common, the chairman lived in one of the houses belonging to the Tudor Admiral who went down with the 'Mary Rose' in 1545. Furthermore, several people present at the meeting were said to have had ancestors living in the village at the time of Monmouth's Rebellion in 1685 and possessed deeds going back to the seventeenth century. To this may be added the fact that in the late twentieth-century Luppit possessed one of the very last old-style country pubs, run in the ancient neighbourly fashion in a room of a house, without so much as a bar counter. Yet the village was otherwise definitely living in the twentieth century: three of the farmers were lately back from Australia and another had just returned from taking his son to the airport to fly there.

The antiquity and 'primitiveness' of the management of Luppit Common turn out to be chimerical, since the present system was devised only when the commoners acquired the land in the 1950s. Its recent creation should be noted. Much of Luppit common had gone compulsorily under the plough during the Second World War. In the 1950s the commoners pressed to recover use of the land and this was agreed, providing some was kept as arable because of the food security concerns of the period. The lord of the manor was obliged to sell the common to the commoners, of whom in 1988 there were about fifty, twice as many as actually came to the meeting that year. Most of them also owned private land. The common rights had been put on record in 1961, when commons were registered nationally, but the commissioner, sitting at Exeter, did not get round to ruling on the claims until 1984. He then threw out some claims of rights to graze sheep and linked those he did accept to each occupier's private holding. This decision snapped the thread of the old arrangement whereby rights attached to every farm; by the 1980s some farms had been sold off as dwelling houses and had lost all agricultural connection.

Typically 160 acres of the total of just under 600 acres of common (in a parish of over 5000 acres) were kept under the plough, but 1988 was the first year when no cereals were grown. The common was used wholly for grazing and haymaking. The management committee wished to keep the ground under grass, improving it with fertiliser and spraying the invasive docks. The hay was baled and the bales drawn for were in lots or half lots.

I attended the meeting of the trustees in 1988 as the guest of the then secretary, H. L. Haynes. He had read economic history before the war under Knoop and Jones (no relation) at Sheffield University and understood my interest. Two dozen people were present at the meeting.

Most were middle aged but there were a few younger farmers. This was a working community: the men were weather beaten and dressed in serviceable clothes, even at an evening meeting. The meeting re-elected its officers, heard the minutes and chairman's and treasurer's reports, all of which business was quite formally carried out. A few thousand pounds were at stake. The meeting then decided, on a vote, to have an annual dinner. The chairman reported on the cost of legal and other actions to try to oust a caravan that had squatted on the common for the past year, the occupant refusing to budge. This was debated in a matter of fact way, with no dislike shown or extraneous comments made but no sympathy for the woman concerned either.

The trustees then stood down in favour of the management committee, which went through much the same business. The agriculturally contentious issues were threefold. First, removing from the chair the power to forbid ploughing next season. The chairman wished to do this to save money. He was stopped because each commoner had the right to ask for a 'ploughed spot', and although it was said no one would insist, rights were thought of as rights, not to be ceded readily. Secondly, forbidding the running of rams on the common. It was pointed out that national legislation forbade this anyhow, at which point one of the younger farmers apologised for having done so, though from his tone he would have liked to do so again. Thirdly, insisting that stock not be run on the after-grass until all grain or hay had been cleared, 'or you wouldn't have any corn or grass left', said one man in an aside. For the moment this restriction depended on a gentlemen's agreement. It was all very calm—what is chauvinistically called very English. No voices were raised, yet at Luppit the age-old tensions about restraining private initiatives within communal farming were evident below the surface.

The Larger Scene

Shifting from the three case studies to consider communal allocation as a whole, the fundamental questions of inception, function and demise can be seen to become speculative as a result of a dearth of documentation. Wet meadows produced the most hay and, until the introduction to drier farms of 'new' fodder crops such as clover and sainfoin, probably the best hay too. Meadow hay was highly advantageous to the keepers of livestock and vital to men with small farms or small holdings which provided few alternative sources of income. The universal indispensability of haymaking and grazing was an incentive for the community to share them out, although at the same time was tempting to anyone who could succeed in privatising the rights. At Hanborough, Oxfordshire, the local farmers acquired over time eight lot meads from the demesne, and although they had to pay Woodstock Manor for their use, they must have thought the grass worth having and that allocating it among themselves by introducing a system of lots was a viable arrangement.[6]

An unusually detailed record of lot-drawing survives from the 1840s and relates to the arrangements at Bampton (actually neighbouring Aston and Cote) a few miles further up the Thames from Yarnton.[7] The author, J. A. Giles, had access to a manuscript book of the seventeenth and early eighteenth centuries containing an account of the system, and even so he found the governance of the common meadow perplexing. This bizarre complexity was not sufficient to preserve the system nearly as late in time as at Yarnton. Aston and Cote Common had certainly been a common for an extended period. To Giles's surprise, an offer to enclose put forward by the lord of the manor in the seventeenth century had largely been spurned. But the lot mead eventually succumbed to the usual onslaught by nineteenth-century enclosers, perhaps prompted by the problems of managing it and the rise of a number of larger proprietors from the ranks of the commoners. Their emergence, it might be guessed, had been encouraged by the increase in market activity by that period, which would have given more scope to the ambitious than the constrained equality when farming was nearer the subsistence level.

Matters were not helped by the fact that the units of area used were as elastic as if they had been devised by Lewis Carroll, who wrote about *Alice* further down the Thames at Port Meadow. In the 1840s about 2000 acres of Aston and Cote were in common field, common meadow or common pasture. The total area was formally divided into 64 yardlands of about 30 acres each, 'but it appears that the yardland is not a fixed measure, but was regulated in great part by the nature of the ground, and varies in different places as much as two or three acres'. The phrase 'nature of the ground' meant relative suitability for growing grass. The remedy was a resort to the ballot, 'upon the principle of justice and equity among all the commoners' so that all might have a chance at the best cuts for their cattle. People in the past were sensitive to such differences, just as they were to local variations in crops with their implications for cutting, stacking and storing grain. Here, then, is a trap for the unwary historian for this is not the only example of variability in areal measurements; tabulating them as if the units were fixed would be totally unwarranted.

Let us ask next about the distribution and chronology of lot meads in general. They had a strong tendency to be located in river valleys, which was where the best natural pastures were found, although the exact position of many is still to be discovered or may never be known. As for chronology, a list of about 100 examples in Berkshire, Gloucestershire, Hampshire and Wiltshire was compiled using the catalogues of county record offices, British History online, and a miscellaneous array of parish histories. Similar lists of places and dates would be possible for other counties, such as Buckinghamshire and Northamptonshire, but this first exercise seems adequate for posing basic questions about the origins, management and demise of the system. Adding further examples produces diminishing returns.

Approximately one-third of the hundred or so lot meads mentioned has a date attached, sometimes more than one date (e.g. 1639, 1692 or 1785, 1817), although almost never with connecting or intervening information. For a few others the attribution is vague (e.g. early seventeenth century or simply seventeenth century). For a larger number we have no more than place names. Many of the references are bare mentions of the existence of a lot mead and very seldom provide additional information about its start, function or cessation. Evidence is thoroughly fragmentary, particularly about the usage of the meadows. Grouping the available dates, a fair number is scattered across the seventeenth century until a cluster appears between 1690 and 1720, and another cluster—the only other cluster—from 1780 to 1820. The majority of references are not dated at all, although from their context they also seem to refer to the seventeenth, eighteenth and early nineteenth centuries. Only a minimum of stragglers occurs later than the 1820s and some authorities refer to Yarnton Mead as an outstanding hangover, or the only one, though that is not quite correct; other late survivors do not however provide such detailed data on management.

There are medieval examples, it is true; consider for instance the ancient nomenclature that Gretton identified at Yarnton. Yet we have no solid grounds for assuming that references in later centuries necessarily derive from as far back. In reality, as Simkhovitch said in 1913, lot drawing can be a late introduction, for example, it was imposed on

Russian peasants as late as the 1830s.[8] And as noted, at Luppit the 'lotting' procedure was devised only in the 1950s! Late introductions raise issues about common attitudes to fairness and awkward puzzles for institutional history. Nor are the present plant assemblages necessarily ancient. The striking displays of fritillaries at places along the Thames may represent a species that has spread only in recent centuries. Nowadays there are an estimated 500,000 individual plants on North Meadow, Cricklade, in a good spring but fritillaries are not mentioned by many early botanists. It may be significant that, whereas the late medieval flora at Fairford, Gloucestershire (down river from Cricklade) resembles the ordinary Thames-side flora, archaeological finds there simply do not include fritillaries.[9]

The sample of lot meads described above is not so much random as haphazard. Nevertheless there is no reason to suppose it is not picking up something real about their history, perhaps phases when communal occupation was widely challenged or came to an end. Care is needed to avoid assuming that two widely separated dates denote long-standing lot meads in a parish as opposed to examples that existed at the first date and were later resurrected, perhaps only temporarily. Absence of evidence is no more than it says, a data vacuum. A reason for uncertainty is that hints appear in occasional local sources to the effect that a little grassland was reserved for some people at the time of enclosure and was then 'lotted', only to fall into private hands later. Alternatively lot meads may simply have been allocated to individuals at enclosure, as possibly happened in Temple Guiting, Gloucestershire.[10] At Ampney St. Peter the lot mead was already in individual hands by 1639. The history of certain lot meads may turn out to have been a fluctuating saga.

Communal Management in General

An attempt to test the possibilities was made by comparing two sets of data for Gloucestershire. Dates attached to lot meads, although not always clear in their meaning, were compared with the dates of enclosure Acts and Awards.[11] Unfortunately, parishes for which there are dated references to both lot meads and Acts and Awards are vanishingly small. Two or three parishes record lot meads only at seventeenth-century dates whereas their enclosures date from the late eighteenth or early nineteenth century. What happened in the interim is not revealed. In two or three other places the mentions of lot meads date from between five years and thirty years before enclosure, raising the possibility that it was enclosure which snuffed them out, i.e. allocated them to private persons. Similarly, in a handful of places lot meads are marked on maps as the property of one or two named individuals, though whether the haying rights were still subject to allocation to others by lot at a fee is unknown.

The impact of Parliamentary enclosure is accordingly hard to assess and a reasonable surmise is that enclosure by agreement at a spectrum of dates was more responsible for ending the lotting system. How voluntary such agreements may have been is moot. What is clear is that the system of communal allocation of haying rights was decaying before it finally ended. At Drayton St. Leonards, Oxfordshire, the lot meads had been subject somewhat enigmatically to 'permanent division' by 1821. Degeneration of the system is more often signalled by the presence of privately owned plots among the strips. Tydalls, consisting of about three acres of Yarnton Mead, was never drawn for because it belonged to an individual and to the rectory of the neighbouring village of Begbroke. At Burford in 1729 two acres in each of four parts of Highmead were likewise exempt; they were called 'laynes' and may have been fenced off.[12] In short, creeping privatisation was hollowing out the system, much as it gradually modified the open arable fields, although completing the process may have taken a very long time.

The original value of lot meads lay in their function as insurance-cum-diversification, meaning that individuals acquired a portfolio of strips differently suited to dry and wet years. This might be of special moment if streams wandered about flat riverside fields and changed the distribution of wet and dry areas. Yarnton Mead was damper than surrounding fields, although not unusably wet in an average year. Inherent dampness could be an advantage in dry years but not strikingly enough to explain why just this and one or two other meadows in the neighbourhood of Oxford survived under lotting systems so late in time. In their very last years the reason seems to have been merely that two or three local people were interested. As to the original purpose, collaborative insurance is held to explain the system of strips characteristic of common field arable farming. Yet in the extensive literature about communal agriculture, and the even greater number of publications about enclosure, the annual reallocation of lot meadow strips is ignored, despite offering an especially concentrated test of the insurance proposition.

Supposing that insurance is the correct explanation of lot drawing, was a further motive equity? Land drained by Vermuyden at Hatfield

Chase early in the seventeenth century was 'cavelled out', i.e. allocated to the investors in equal shares. Such a division, impersonally arrived at, might have been a means of avoiding disputes. Village people wished to avoid them too. They lived close to their neighbours and needed their help; families were often intermarried; and while some people did shift from place to place, especially in the nineteenth century, most inhabitants would not move away in the short term. Children and young people would grow up together. In other words there was a sense of community which may have made sharing seem automatically 'fair' to most. But although proximity makes neighbours it does not necessarily make good neighbours. Sharing might have been altruistic only in part. If a family with a legal settlement fell on the poor rates then the parish would have to keep them and it might seem more efficient to encourage them to feed themselves at however a low level. Farmers were not known for their charity, as nineteenth-century commentators as different as prominent rural diarists like Peter Hawker in Hampshire and Cecil Torr in Devon were quick to point out. All told, the avoidance of disputes seems likely to have been the leading motive for strip-farming and for lotting.

Against the advantages of diversification-cum-insurance were certain disadvantages. One was the inflexibility enjoined by communal decisions. At Yarnton the auction and ceremony were scheduled for the same days in July every year, whatever the weather. Already in 1789 William Marshall was railing against the fixed dates when common grazings were thrown open at various places in the Vale of Gloucester, calling this an injudicious 'RELICK OF ANCIENT CUSTOMS'.[13] Moreover, under a lottage system no individual on his own was likely to spread manure because he could not know which strips would fall to him next season.

Altering the species mix was also difficult, although farmers did try to halt the spread of dock, thistles and garlic, the last of which tainted the milk. Whether any species were actively encouraged is not known, leaving aside aspirational statements in the improving literature; possibly butter from rich mixtures of herbage was actually preferred. The hay had been more valued when horses were the means of transport and farm traction. Modern studies do not seem conclusive about how nutritious mixed herbage was, partly because the mix varies from place to place and year to year. Although it is obvious that commoners knew perfectly well about the fodder crops and artificial grasses that had been introduced from the seventeenth century on, the communal system was too rigid to permit much by way of demonstration effects from neighbouring farms in severalty.

The demise of lot meads as a whole is something of a black box. How far it came about through voluntary agreement or was imposed by 'taste leaders' or via top-down enclosures is unfathomable. The closest we can come to the forces at work is by considering such evidence as there is about the experiences of the common arable fields. On the negative side, take the big Oakridge Common in Gloucestershire, on which the poor had fed their donkeys. When it was enclosed all they received were small plots well away from their houses, which they were obliged (but could not afford) to fence, so they accepted a load of firewood at Christmas in lieu, and after a few years even that stopped.[14] They were left with the dubious satisfaction of running alongside the lord of the manor's carriage at election times, chanting, 'who stole the donkey's dinner?'

Mary Sturge Gretton, who was extremely well informed and thoughtful, described the enclosure of Burford, Oxfordshire, not far distant, as a more benign affair.[15] Enclosure of the old common arable replaced it with the walled fields and pastures of the subsequently familiar Cotswold landscape. That was in 1773 and 1795. Of 1100 acres only 300 went to independent farmers, the remainder going to that descendant of Cromwellians, John Lenthall, and the Impropriator of the Rectorial tithes. Small farmers working on their own land almost disappeared. But Mrs. Gretton pointed out that new springs of prosperity in the sixteenth century had already drawn yeoman families into the town; they had *willingly* parted with their own land while the Lenthalls and one other family were amassing their estates. This made way for a new class of farmers who were merely tenants. Before we dismiss the voluntary element in this, consider how hard and uncertain the working farmer's life was, and how much preferable life behind the shop counter may have seemed.

The processes were generalised by J. B. Phrear in an 1889 article about Braunton Great Field, Devon.[16] At Braunton the land is still to be seen laid out in strips, although it is no longer farmed in common. Phrear made the Great Field his starting point for a general discussion of agrarian change. Medieval villeins, as he conceived it, had obtained their shares in the common ground by paying a one-off fine or a rent in money to the lord of the manor. Then, strip by strip, the shares of the smaller holders passed into the hands of the larger ones until all belonged to one owner and were consolidated, then enclosed with a hedge. 'And it may perhaps be conjectured, without much risk of substantial error', continues Phrear, 'that the transition from the manorial system of Anglo-Saxon times to the landlord and tenant system which prevails now in England, was brought about by some such steps as these'. This makes it seem natural and automatic, a one-way street, and indeed studies by economists do show how fast greedy (or entrepreneurial?) individuals are able to amass property. But in that case, why did communal systems take so long to disappear?

Instances did occur where local farmers readily agreed among themselves about enclosing. In the 1580s twelve tenants at Monxton, Hampshire, undertook a joint enclosure and afterwards wrote with consummate effrontery to the lord of the manor, which was King's College, Cambridge, stating, 'Honoured sirs, Without your consent and privitie, we have presumed to enclose our Cowdowne, not doubting but (when you shall understand it) you will easily approve our doings, as being an improoviment of the soyle and a profit, both to your selves and us'.[17] One of these ingenuous improvers was an Edward Noyse, almost certainly an ancestor of my wife.

Processes that caused some men to lose their land or rights, or to sell out and descend into wage labour, were operating so constantly as to seem inevitable in retrospect. Every so often a personal crisis, maybe a health emergency, might mean that someone faced a sudden demand for cash beyond what could be met by borrowing within the family or from neighbours. Another reason for relinquishing rights may have been emigration, with the rights sold to wealthier inhabitants.[18] Giving up one's holding or equivalent rights would seem an extreme response, like selling the family silver, but the history of gradually increasing social inequality suggests that it happened often enough. An example of unequal arrangements that presumably started equitably comes from Aston and Cote where a majority of the farmers had only one-half or even one-quarter of a yardland in the common fields but some individuals possessed ten or eleven yardlands.[19] As Mary Sturge Gretton implied, expanded job opportunities outside agriculture may have made selling out seem a little easier as time went by or at least provided anyone who gave up small-scale farming with openings in wage labour.

Enclosed parishes pushed out those who could not keep their heads above water in the advancing market economy. The process was long standing and overall rather slow. 'The cottiers, who earn their living in work connected with the plough', wrote a minister in the early seventeenth century, 'are now cast adrift and wander towards the nearest open-field villages – thus it comes about that the open-field towns have about double the number of cottiers they used to have, and are hard put to it to provide for the poor which the enclosures have made'.[20]

Efficiency and equity tugged at one another in common field villages. Day-to-day debates were over transaction costs (the expenses of agreeing about ordinary management, let alone innovation) and exclusion costs (how to keep other people's beasts out of one's crops). There was also a journey-to-work problem. Historians are fond of stressing how inconvenient it must have been for the farmer to move among his scattered strips and in this respect their instincts seem correct. Mr. Rose, bailiff at Laxton remarked to me in the 1950s, 'you're always on t'road in Laxton'.

Disputes about the means of maintaining soil fertility were endless. What was to the advantage of the individual, such as taking turves and manure from the common to put on his own land, was to the disadvantage of his fellows. Local histories are full of complaints about this, rectified at the courts leet. As we saw at Yarnton, the timing of operations was another potential source of friction. In the seventeenth and eighteenth centuries people were starting to risk growing crops of their own in the common arable fields, especially 'new' fodder crops like sainfoin, which led to disputes about the date at which cattle might be let in to graze. Side payments were required. The opening to cattle was called 'the breach'. At Calne, Wiltshire, in 1751, 'Robert Capp by wetness of season not having been able to cut his wheat...' was obliged to compensate his fellow commoners for deferring the breach.[21]

Exclusion costs, therefore, could be high. Boundaries were always important when land was the means of subsistence. 'Most medieval Englishmen lived in constant fear of trespass', according to John R. Stilgoe.[22] This fear continued long after the Middle Ages, because in the fluidity of open fields the boundaries could be marked only sketchily, using shallow ditches and corner stones, or temporarily in the standing hay by 'treads' like those run to mark out the strips at Yarnton. The danger was of animals getting loose in the crops. The plea of Cressy Dymock in a letter to Samuel Hartlib, published in 1653, was, 'I have observed the proneness most of Cattle & Poultry to break into forbidden places...[pigs, rabbits and pigeons] are most inclined to and frequent actors of mischief, and that so great, that men dare hardly consider it seriously, but let it passe to avoid vexation'.[23] Dymock's conclusion was that, 'the evill Contrivance and Inter-mixture of wayes and Interests in most places of *England* is a special reason...' why improvers have no place secure against the carelessness and wickedness of neighbours who let in their beasts and fowls.

Pigs were bad offenders. They could not easily be seen in the crops and poor people covertly drove them in. The pigs would 'but play bo-peep' and if the farmer tried to oust them without a dog they would trample still more. Cowherds ('Little Boy Blue, come blow your horn') were needed, 'to restrain the cows from springing into the arable fields', as Pehr Kalm, a Swedish visitor to Buckinghamshire, wrote in 1744.[24] All this reinforced the interest of enterprising farmers in gathering their scattered lands into a single piece, preferably close to their homesteads, and if possible ring-fenced—in short, to enclose and privatise. Lot meads were not exempt from this development.

Lot meads were a special case of common field agriculture and their main features can now be summed up. They were not ancient in all cases and their history may have been recurrent. The most plausible explanation of their purpose is that a lottery was the fairest means of diversifying the portfolios of the participants. It offered them cover against extremes of weather and the differing potential productivity of the land. To modern eyes the divergences might seem almost trivial but they were significant, even crucial, to men who owned few livestock and whose holdings were small. Equity was thus likely to have been a major consideration but avoiding disputes through what amounted to annual balloting may have been even more important.

A central question is whether lotting was inflexible or not. The tenor of the literature about common field farming is definitely that it was slow to amend. It is true—it was obvious at Yarnton in the 1960s—that fixed dates of mowing meant risking bad weather at haymaking, something which the occupants of holdings in severalty could hope to avoid. Additionally, there were problems of the 'journey-to-work' variety and a constant need to exclude roaming animals. Strip farming of all types did not completely preclude innovation but undoubtedly hampered it; an agreed date for mowing was especially hard to circumvent, so visible was any transgression.

Against these defects, elements of flexibility did enter, as is shown in Aston and Cote. It manifested itself in the informal, 'out of system' exchanges designed to reduce journey-to-work costs or the awkwardness of scything very narrow strips. Pragmatism was also revealed by the elasticity of areal measurements, where yardlands might vary considerably in adjustment to the different expected productivity of pieces of land.

Measurement outside lot meads sometimes displayed comparable pragmatism, which once again cautions against taking written numerical records at face value. Nevertheless a degree of inflexibility was probably dominant. Despite this, emerging unequal power structures in the common fields enabled the richer farmers to argue successfully for change. They could band together in favour of enclosing, which was a means of cutting the Gordian Knot. This presumably applied to lot meads too, if some men had amassed rights over a majority of the strips. The creeping or intermittent decay of the system seems to have owed most to internal change.

Notes

1. R. H. Gretton, 'Lot-Meadow Customs at Yarnton, Oxon', *Economic Journal* XX (1910), pp. 38–45, and 'Historical Notes on the Lot-Meadow Customs at Yarnton, Oxon', *Economic Journal* XXII (1912), pp. 58–62. Yarnton Mead has long received intermittent notice, latterly from botanists. Descriptions have been offered of the lot-drawing, the balls drawn as lots, and the vegetation of the meadow, but analyses of the motives underlying the system, here and in other lot meads, are absent or scarce. See however Mrs. Bryan Stapleton (M. H. A. Stapleton), *Three Oxfordshire Parishes* (Oxford: Oxford Historical Society/Clarendon Press, 1893) and the brief account of visits in 1931 to three remaining common-field sites by C. D. Linnell, 'Survivals of Pre-Enclosure England', in J. W. Robertson Scott, *The Countryman Book* (London: Odhams, 1948), pp. 267–271. A well-informed first-hand account of the Yarnton arrangements is F. J. Charlett's statutory declaration before Albert J. Henman at Woodstock, 30th Sept. 1946. Charlett had attended every lot-drawing for forty-one years until 1946.
2. H. Baker, 'Alluvial Meadows: A Comparative Study of Grazed and Mown Meadows', *Journal of Ecology* XXV (1937), pp. 408–420.
3. Gretton came from Burford, Oxfordshire, where lot-drawing dated from at least the fourteenth century and continued at least into the eighteenth century. It was practised at nearby Taynton in the seventeenth century and may have ended in Temple Guiting only at the enclosure of 1804. Anthea Jones, *The Cotswolds* (Chichester: Phillimore, 1994), pp. 81–83.
4. J. A. Giles, *History of Bampton* (Bampton: Privately printed, reprint of second edition of 1848), Section 20: Of the system of farming which prevails at Aston and Cote, Rights of Common, &c., pp. 75–82.
5. The minute books from 1735 to 1921 are housed in the Wiltshire and Swindon Record Office.

6. Stephen Braybrooke-Tucker, *Hanborough* (Bloomington, IN: AuthorHouse, 2012), pp. 87–89.
7. Giles, *Bampton*, p. 80.
8. V. G. Simkhovitch, 'Hay and History', *Political Science Quarterly* XXVIII (1913), pp. 398–399.
9. Mark Robinson, Archaeology of Hay Meadows, Lecture to Hay Meadow Workshop, Wytham, Oxford, 23 June 2007.
10. Anthea Jones, *The Cotswolds*, p. 82.
11. Enclosure Acts and Awards were taken from W. E. Tate, 'Gloucestershire Enclosure Acts and Awards', *Transactions of the Bristol and Gloucestershire Archaeological Society* 64 (1942), pp. 1–70.
12. Anthea Jones, *The Cotswolds*, p. 81.
13. William Marshall, *The Rural Economy of Glocestershire* (Stroud: Nonsuch, 2005 [1789]), p. 112.
14. Norman Jewson, *By Chance I did Rove* (Privately printed, 1973), p. 39.
15. Mary Sturge Gretton, *Burford Past and Present* (London: Faber & Faber, 1944 edition), pp. 33–34. Comparable indications appear in Crawley, Hampshire. N. S. B. and E. C. Gras, *The Economic and Social History of an English Village* (Cambridge, MA: Harvard University Press, 1930), pp. 124–125.
16. J. B. Phrear, 'Note on "Braunton Great Field"', *Trans. Devon Association* XXL (1889), pp. 201–204.
17. Diana K. Coldicott, *Monxton: A Hampshire Village History* (Andover: Privately printed, 1998), p. 32. Interestingly, in 1620 King's College was actually putting intense pressure on its tenants (at Stour Provost, Dorset) to accept enclosure and land reallocation. J. H. Bettey, *Discover Dorset: Farming* (Wimborne: The Dovecote Press, 2000), p. 25.
18. As at Crawley, although strictly this does not refer to lot meads but to cottages and other real estate, the principle seems the same. Gras and Gras, *English Village*, pp. 124–129. Windfalls might produce the opposite result when servants received bequests from their employers. At Amport, Hampshire, by the 1899 will of the Marquess of Winchester, eight of the staff received £500 each and a housekeeper and kitchen-maid £350 each. Given that Bridge Cottage (close to Amport Green) sold for £90 in 1923, a kitchen-maid might become a person of property as a result of this legacy. Amport.Archives@gmail.com. Retrieved 19 July 2020. I am indebted to Shaun Kenaelly for this reference.
19. Giles, *Bampton*, p. 76.
20. A. C. Chibnall, *Sherington: Fiefs and Fields of a Buckinghamshire Village* (Cambridge: Cambridge University Press, 1965), p. 200.
21. A. E. W. Marsh, *A History of the Borough and Town of Calne* (Calne: Robert S. Heath, 1903), p. 85.

22. John R. Stilgoe, 'Jack o'lanterns to Surveyors', *Environmental Review* I (1976), p. 19.
23. Samuel Hartlib, *A Discoverie for Division or Setting Out of Land* (London: Privately printed, 1653), pp. 3–4. I have elided some redundant phrases. The pigeons would have come mostly from private dovecotes.
24. Quoted by Vicars Bell, *To Meet Mr. Ellis: Little Gaddesden in the Eighteenth Century* (London: Faber & Faber 1956), p. 116.

CHAPTER 8

Drove Roads

The Flow of Livestock

Southern England used to be patterned with drove-ways for sheep and cattle like the grain in timber. The main routes ran from west or north-west to east, converging on fairs where farmers from a ring of counties around London bought animals bred in Wales or the West Country to fatten closer to the metropolitan market. This tied together the historic sheep economy. My father, who was a boy before the First World War, told me that during late summers or dry autumns, flocks of sheep kicked up white clouds across the chalk lands. The dust powdered the hawthorn hedges and could be seen for miles. What he saw were flocks from the West Country converging on England's greatest and most famous sheep fair at Weyhill.[1] In the same Edwardian period, Bob Wilson's head drover at Bishopstone, North Wiltshire, once had twelve flocks in view, also trailing white clouds. The Bishopstone flocks were heading eastwards along the Ridgeway towards another great fair, at East Ilsley, Berkshire. It has all gone now, at least as far as moving animals on the hoof is concerned.

Cross-country movements of livestock were reported in the Middle Ages and by the early seventeenth century were known to have headed from Scotland and Wales towards London in sizeable numbers. Such commentaries as there are nevertheless imply that long-distance droving

E. L. Jones, *Landscape History and Rural Society in Southern England*, Palgrave Studies in Economic History,
https://doi.org/10.1007/978-3-030-68616-1_8

became more intensive during what has been termed the London cattle- and sheep-droving period.[2] This began in the late seventeenth century, and flourished during the eighteenth- and early nineteenth centuries in step with the growth of London's population. It fended off the railway threat and only gradually subsided, until between the two world wars it died right off. Across central southern England a rise in traffic during the economic expansion of the late seventeenth-century certainly accords with other signs, notably of the emergence of an apparatus to support travel. This included the development of a little port town at Lechlade which was the effective head of navigation on the Thames (and through which animals that had come down the 'Welsh Way' via Fairford, Gloucestershire, were conjecturally driven en masse towards the Ridgeway). In similar fashion, Birdlip on the Cotswold escarpment was established as a village of inns, stables and blacksmiths' forges to service an expanding trade.

Convergent Routes

Escaping the spell of the old droving routes would have been hard in the south of England. Droves converged and became so dense in the Andover district, where I grew up, that overlooking them was scarcely possible. Andover was a mere three miles from Weyhill, the origins of which have been lost, as the saying goes, in the mists of history. The fair was held without a founding charter and continued through the years for no better warrant than that it always had.[3] Furthermore Weyhill was only two miles from an upstart rival, Appleshaw fair, while the borough of Andover had come to hold its own sheep fair too.

In my boyhood the great fairs were closing but in the early 1960s I did experience the fair at Wilton, near Salisbury, at a low point in the sheep business. Unlike so many fairs, Wilton is one that has managed to keep going. In the early nineteenth century 40,000 sheep were sold there every year and A. G. Street, writing in 1931, had seen 70,000 sheep penned and sold in a single day.[4] A still creditable array of 12,000 breeding ewes, store lambs and rams were on offer at the biggest of Wilton's four sales in the 2019 year. This may not match the historic scale, nor are the animals any longer driven in on foot, but is a lively reminder of an age-old trade.

Despite its superior scale and longevity, the Weyhill market did not survive. An interesting history could not save it: Chaucer had referred to

Weyhill fair (which his granddaughter owned) and Hardy had transposed to it a real incident when a man put his wife up for sale in Andover. The large-scale traffic to the fair tempted neighbouring settlements strategically placed on drove routes to try to intercept the flows. They hoped to establish fairs themselves, if they could secure charters to do so, for the sake of the profit from letting coops in which to display animals for sale. Central government was willing to oblige, at a price. Weyhill outlasted the surrounding challengers, although it was unable to prevent some of the trade being usurped, until at last it closed. Its final show of sheep and cattle was held in 1957.

Most of the farmers who sold their livestock at Andover came from within a ten- or fifteen-mile radius but a strong contingent came from further afield, in Wiltshire and Dorset. A share of this west-east trade, which was what underpinned Weyhill, was worth trying to acquire. Late seventeenth-century Andover overinvested in a complicated series of legal actions aiming to capture the business of Weyhill fair for itself. Other towns and districts tried to resist Andover's expansionary efforts.[5] But the Borough was well advised to establish a fair of its own in order to make money from letting fees. Accounts of the town's sheep fair and coop-lettings survive for 1818–1834, when the annual receipts fluctuated between £72 and £182.[6]

The village of Ragged Appleshaw, which lay on a north-south drove and was even closer to Weyhill than was Andover, secured the right to hold two fairs in 1658 in similar rivalry with the great fair. Appleshaw cheekily held its fairs on the Friday and Saturday before the week of Weyhill fair; it sold 15,000 sheep in 1801 and had sold more previously. While Appleshaw's move was profitable, it did not succeed in matching Weyhill, where at its height in the first years of the nineteenth century 140,000 sheep were supposedly sold, although the *Hampshire Chronicle* claimed that a penning of 90,000 in 1816 was the largest for over thirty years.[7] In the same district during the eighteenth century, the father of the millionaire Victorian haberdasher, James Morrison, had built pens for 3000 sheep at the George Inn, Middle Wallop (five miles west of Andover), together with accommodation for drovers on their way from Scotland to Salisbury, Stockbridge and Wilton Fairs.[8] The Morrisons had arrived from Scotland and were rumoured to be of drover stock. They had come to a shrewd location for profiting from the transit of livestock.

Reaching the Ridgeway

Cross-currents there were because an array of local markets existed (naval demand at Portsmouth was important) but the metropolitan demand for meat was the chief impetus orienting long-distance droving.[9] Vast streams of livestock converged on London via complex routes and intermediate fairs, but essentially from north and west. One stream, primarily cattle, came through East Anglia from Scotland, and has been well studied. The other was multiple, consisting of West Country and Welsh sheep and cattle either moving to Weyhill or travelling eastwards to East Ilsley along the northern rim of the chalk lands, on the most famous of all Ridgeways.[10] Some of this stream was made up of cattle from Wales which in the late eighteenth century were carried across the Severn from Beachley on the 'Great Boat' of the Aust ferry (Much later replaced by the first Severn Bridge). Young steers and wether sheep, already well-fed in summer on the Black Mountains, arrived at the ferry via Brecon and Abergavenny. They were directed from Aust towards the Ridgeway and from there to East Ilsley en route to the London area.

Sales at sheep fairs fluctuated from year to year and in this respect East Ilsley was no exception. The numbers supposed to have been sold there vary according to authority; Pigot's Directory of 1830 claimed that 25,000 sheep had been sold in a day and that the annual total ran to over 250,000. They were principally bought by farmers from Hertfordshire and Buckinghamshire, who fattened them for the London market, just as they fattened animals from many districts far to their west and north.

Unlike Weyhill, the date of East Ilsley's inception is known: in 1222 Henry III had granted the right to hold a market. Official recognition nevertheless failed to obviate the constant squabbling between settlements anxious to cash in on the trade, notably the protests of Wallingford that East Ilsley would harm its own market. This did not succeed in stifling Ilsley, which at least in later times drew large numbers of sheep from within a twenty-five-mile radius (hence including the home turf of the aggrieved Wallingford folk) and not merely animals arriving via the great trek from Wales. Some local animals were brought by drovers from Abingdon along the pronounced causeway at Steventon, explicitly described as a dry route in an otherwise boggy countryside.

The Ambiguity of Droves

Wherever possible the long-distance drovers used ridgeways, of which the Wiltshire-Berkshire-Oxfordshire Ridgeway is the type specimen. Ridgeways occurred in many counties, indeed in many countries. People had always chosen to travel over dry, fairly bare, high ground (long ago cleared for sheep farming) when the low vale country continued to be thickly wooded. So it is said: we need not wholly disagree but should add that even after much of England was substantially disforested the country remained a wet environment inhospitable to travel and the clay vales especially so. As Tacitus had said in *Agricola*, there was 'extreme moistness of land and sky'. The land stayed ill-drained long after the main phases of woodland clearance, until the field drainage of Victorian times.

Enthusiasts have traced drove roads all across the landscape, not merely those on ridges. They have found markers such as the isolated stands of three pine trees which sign-posted the routes, large dew ponds, and handily located pubs. Sometimes they have found picturesque descriptions of drovers and occasional counts of animals passing along, although more often only allusions to the movements. Several books and articles have been written about the more dramatic aspects of droving, especially from deep in Wales right across to Kent, or from as far as Skye through East Anglia to the markets ringing London. Details of lesser movements are hard to come by; those which are available tend to be ad hoc mentions.

England is full of potentially suitable green lanes, tracks and droves, intriguing but usually impossible to date. The mute testimony of features in the landscape represents form but not function. Hurst in *Sheep in the Cotswolds* urges that, 'an archaeological approach… is of little use for detailed investigation of even the most basic themes of food supply, or the supply of perishable raw materials such as wool, or cloth'.[11] This reminds us that the physical presence of drove-ways, or supposed drove-ways, proves little. It is silent about usage. Even so, hints about former agricultural uses are sometimes prompted by the landscape and can be fleshed out from manuscript or printed sources which might otherwise be passed over. The reverse approach may be helpful too, when documentary or literary sources suggest searching for ground evidence the distribution of which may make obscurities in the written word more comprehensible. For all sorts of topics, evidence is too sparse to ignore clues of any type.

Their presence on the ground is no guarantee that the innumerable routes were in use at the same time. Little is known about their function at any given date. Charting them would require an impossibly vast array of maps and give the unfortunate impression that they were simultaneous features of long-distance networks. They were employed for miscellaneous purposes, often locally when farmers moved stock from one end of their farm to the other or villagers drove cattle to graze on the common. Uncertainty as to which tracks were genuine long-distance droves arises from the fact that, if two or more villages put animals on the grazing land which they held in common, the routes might be indistinguishable from throughways. Tracks also connected villages to detached parcels of meadow and common over which they had rights, for instance, in the grasslands of the Thames where surprisingly distant parishes sometimes owned land. These meadows might be reached from any number of points, requiring and leaving traces of all sorts of lanes.

The Thames valley is a wet country and entry could be impeded by marshy stretches, made passable by building causeways. Causeways occur in a wide scatter across the country, testifying to enormous labour with pick and shovel, although we cannot know how many seasons (probably winter months when farm work was at low ebb) were taken to construct them. Cattle can of course swim; they were roped head to tail and made to swim from Skye to the Scottish mainland and along with sheep from Anglesea to the Welsh mainland. But in the tamer English lowlands such drastic measures were unnecessary: the problem there was to prevent livestock from becoming bogged down in field bottoms and rain-filled sloughs, hence the construction of local causeways.

Managing Movements

In addition to the complex of fairs at Weyhill, Appleshaw and Andover, two other two places where I have lived for longish spells are Mortimer, Berkshire, along what is known as the 'Welshman's Way', and Fairford, Gloucestershire, where the 'Welsh Way' arrives from mid-Wales. Even more explicit traces of the Welsh survive, such as the inscription on the former inn called Drovers' House at Stockbridge, where the River Test is crossed between Salisbury and Winchester, which reads in translation, 'Season's Hay, Rich Grass, Good Ale and Sound Sleep'. There was also a major droving route across the Midlands called the Welsh Road. These routes were ancient and long travelled by Welsh drovers, although the 'Welsh' label was sometimes applied only in the nineteenth century when the Welsh were the last people to keep to green lanes as a means of circumventing tolls on the roads and charges on the railways.[12]

On maps the long-distance droves may seem as though they knit together strings of towns but on a closer look the green lanes tend to miss the settlements and above all skirt around their central areas. Jostling flocks of sheep would have been a messy nuisance to tradesmen and their customers, and measures were needed to minimise the problem. A friend of mine, taken on during the Second World War as a young shop girl at a drapery in Sleaford, Lincolnshire, was told her first task each market day would be to put up the 'sheep sheets'. Mystified at first, she learned that sheets were hung around the lower level of the windows to prevent the sheep from seeing their reflection, panicking and breaking the glass.

Managing the flocks, besides herds of cattle and horses, was difficult enough to focus attention in many small towns. In the Gloucestershire and Oxfordshire Cotswolds sheep were funnelled to local markets through narrow lanes or alleyways called tures, which controlled movement and made the animals easier to count. Elsewhere (in Andover for instance) inn yards which led to and from market places fulfilled the same purpose. These features, and sometimes an inner lane actually called ture road, the tuer, or some variant, remain as witnesses to an everyday business now almost forgotten.

According to Natural England 'ture' is a local word known in 1990 to fewer than ten people in the country.[13] But ture has two connotations, one referring to long hillside corridors through which sheep grazed on dry hill-top pastures were taken down to water, the other (which concerns us here) denoting short urban passageways to make driving sheep in and out of market places easier and reduce any damage they might do. In the latter sense, ture (tuer, etc.) was a management device adopted to deal with the local trade of market towns. Long-distance drovers tried to avoid the cramped medieval centres of towns and villages entirely; if they had to come close, they used the back lanes that commonly ran behind houses and cottages. All the testimony is that drovers kept away from both village streets and toll roads. They threaded their way between what they saw as hazards.

Modern maps show a variety of routes but to discover how and when each and everyone was actually used it is necessary to rely on what is pejoratively called hearsay. In truth, hearsay amounts to oral testimony (one might say 'witness statements') by people who had no reason to embroider what they said and were old enough to have learned the topography of where they grew up from parents and grandparents who had grown up there too. Thus inhabitants of Fairford have told me about two small fields used to rest livestock which had come down the Welsh Way en route to Kempsford and headed from there to Highworth market or beyond; one of the fields was quaintly called Timbuctoo. I could not have learned this from documentary sources nor deduced the purpose of these little fields just by looking at them.

Volumes of Traffic

Number of animals on the hoof or penned at market are typically stated in thousands, but suspicious rounding and many discrepancies hint they are not always to be taken literally. Gwyn Williams cites an eighteenth-century estimate, which may have been more credible than most, to the effect that 15,000 cattle and 5000 pigs were exported annually from Anglesea but even his source gave up with sheep and merely said they went in their thousands.[14] In principle, market tolls give precise counts but oscillate so much from year to year as to provide only a rough index of the average traffic. The fluctuating numbers mirror variations in the prosperity of sheep farming as a whole, not simply droving, and can be shown to pick out years of boom and slump. The longer trends make it clear however that for three or four centuries the droving trade which supplied the fairs was sizeable. Defoe, who had been bewildered in the first years of the eighteenth century by the large number of sheep he saw at Weyhill but could get no credible estimate from a local grazier, retreated to saying that, 't'is sufficient to note, that there are a prodigious quantity of sheep sold here'.[15] It is tempting to leave it at that.

A more interesting point may be that long-distance droving was not after all as large as the figures given for sales at regional fairs imply, which would be too easy to accept at face value. Much of the supply is said to have come from farms within, say, twenty-five miles of even the biggest fairs. Sheep farming was a major activity in the countryside, from which farmers came to the fairs for the animals they needed to replenish mortality among their folding flocks, whose manuring was central to their rotations. Thomas Mozley, rector of Cholderton, Wiltshire (seven miles from Weyhill) in the 1830s and 1840s, noted that the farmers in his parish only half-reared their sheep until they were solicited by the fatteners for the London market.[16] The romance of long-distance droving has been better than local husbandry practices at catching the attention of both contemporaries and historians.

Interpretive problems arise because the volume and above all the value of the trade eludes the scholar, beyond the broad conclusion that usage increased with the growth of London until droving was challenged by alternative modes of transport. The only recourse is reports of the number of animals on offer at the fairs and markets, although they are at best spotty or blurry. Photographs of animals actually on the hoof in long-distance transit are astonishingly scarce. Nineteenth-century photographs do exist of the fair at East Ilsley.[17] Unfortunately the photographs are no more than incidental evidence because they cannot reveal what proportion of the animals shown was actually from local farms, as opposed to far-off Wales or Devon and Cornwall. In any case photography can capture only the twilight era of droving.

South-East from the Cotswolds

Reports do exist of the livestock movements across the Cotswolds, through the Upper Thames area, and along the Ridgeway. Black cattle from Wales came via Gloucester and the Cotswolds to be fed ready for the Christmas market at Smithfield in London.[18] The drovers, or local entrepreneurs cashing in on the traffic, had a halting-place at Hucclecote, between Gloucester and the foot of Birdlip Hill, and another holding paddock in Birdlip village. A little further on, down the Cotswold dip-slope, the mid-eighteenth century rector of Barnsley let the rectory ground to drovers halting overnight, charging enough to make a pleasing £5.17s.0d in 1767. He was cutting out the inn at Ready Token, which was the regular stopping place. Still further on, farmers from the Upper Thames area fetched cattle on their own account, buying them at Hereford October Fair and fattening them on pastures in the Upper Thames district, such as Walcot, Clattinger and Kelmscott. Specialist dealers in sheep also visited western districts in search of animals to sell to the fatteners in counties close to London, in other words to cut out those other middlemen who operated the west-east trade through the fairs.

Sheep were the most numerous of the animals passing through. They came, with additional cattle and some ponies, down the Welsh Way to Fairford. There is a degree of uncertainty about where they went next, indeed the movements were probably various. The animals arriving at Fairford down the dry Cotswolds have always been supposed to have passed on through Lechlade and crossed the Vale of White Horse to Wanborough, just below the Ridgeway in Wiltshire, by whichever route minimised the distance they had to traverse over the wet clays. As I have already implied, it is by no means certain that the main route really was from Fairford to Lechlade, although livestock do seem to have passed right through the former. Irish copper coins have been found at the site of the ford across the Coln, close to the present Fairford town bridge, and it has been speculated they were 'luck pennies' thrown in by drovers conveying Irish cattle from Bristol to London.[19] Moving herds of cattle and flocks of sheep over Fairford's narrow bridges and through its cramped streets (tortuous ever since late medieval 'road capture' had diverted the London Road at the town bridge) was not an enviable task or likely to be popular with residents.[20] Lechlade, too, would have been unhappy about a big passage of animals.

Details of local topography are needed even to speculate about droving at this level, emphasising just how knowledge about the trade can break up on examination. A possibility is that some of the animals were manoeuvred along the edge of Fairford common and via Totterdown lane to the good pastures on the Thames at Kempsford, and then through Hannington to Highworth, which town was not unimportant as a local market. From there, instead of through Lechlade, the Vale of White Horse might have been crossed to the Ridgeway above Wanborough. Perhaps there was more than one route but it is all conjectural and emphasises the difficulty of determining the usage of droving routes in the absence of documentation.

The Ridgeway was an ancient and important route for both sheep and cattle that had been brought from a great distance.[21] A writer in 1861 noted that cattle were being driven along drove-ways right from Anglesey as far as Kent, the merit of such routes being that they charged no tolls and almost entirely avoided bridge crossings (choke-points which were hard for the drovers to master).[22] In the nineteenth century Welsh drovers were turning their attention more to sheep, which perhaps helps reinforce the subsequent impression of agricultural fairs in south-central England as predominantly sheep fairs. But the fairs sold many other commodities, for example, Weyhill was a leading market for hops, which came east-west in the opposite direction from the main routes of sheep. The fairs may have gone on looking chaotically busy however many or few 'foreign' sheep were sold.

Tailpiece

At Ilsley, which was probably representative of south-country fairs, sales peaked in 1851 and 1861. This corresponded to the 'Golden Age' of English farming, after which a decline started. It was not the railways that killed off the droving trade, despite a shift of business in their direction. Railheads were fairly widely spaced and sheep still needed to be driven to the station. What administered the death-blow to droving was the lorry, a vehicle that began its civilian life in the 1920s as a tray placed on top of a war-surplus chassis. Lorries were flexible and could fetch animals at the farm-gate for delivery anywhere. Drovers continued to operate, tenuously, until just after the First World War but subsequently their occupation failed. The Berkshire Ridgeway sank out of commercial importance, used only by local farm vehicles and leisure walkers.

Notes

1. For Weyhill see especially Robert Clutterbuck, 'Notes on the Fair at Weyhill', *Hampshire Field Club and Archaeological Society, Proceedings* 3 Part 1 (1895), pp. 127–142.
2. Among a number of books on routes the most thoughtful and engaging remains C. Cochrane, *The Lost Roads of Wessex* (London: Pan Books, 1972). See p. 98 on the expansion of cattle-droving.
3. Holding a fair and charging fees for erecting sheep-coops or fair booths were apparently separate rights. The plots of land in use changed over time without interrupting the fair's continuity.
4. A. G. Street, *Farmer's Glory* (London: Faber & Faber, 1937), p. 32.
5. For example, the 1672–1673 petition to the Lord Chancellor by gentry and graziers in north-west Wiltshire, protesting at the establishment of a new market at Andover. Goddard Collection, 1337, Swindon Public Library. Clutterbuck, 'Notes', cites several other petitions.
6. Town (Sheep) Fair Accounts, 1818–1834, Andover Borough Archives.
7. 1816 saw many sales of other live and dead stock, besides the disposal of actual farms, reflecting the agricultural slump of 'the year without a summer'. A large offer of sheep in 1816 was likely to have been a sign of distress as much as of a boom in the industry.
8. Dorothy Beresford, *Nether Wallop in Hampshire* (Privately printed, 1989 edition), p. 5; Caroline Dakers, 'James Morrison (1789–1857), "Napoleon of Shopkeepers", Millionaire Haberdasher, Modern Entrepreneur', in Christopher Breward and Caroline Evans (eds.), *Fashion and Modernity* (Oxford: Berg, 2005).

9. The most informative article on droving is Cynthia Brown, 'Drovers, Cattle and Dung: The Long Trail from Scotland to London', *Proceedings of the Suffolk Institute* xxxviii/4 (1996), pp. 428–441. See also C. Skeel, 'The Cattle Trade Between Wales and England from the Fifteenth to the Nineteenth Centuries', *Transactions Royal Historical Society 4th Series* IX (1926), pp. 136–148; G. E. Fussell and C. Goodman, 'Eighteenth Century Traffic in Livestock', *Economic History* III/2 (1936), pp. 223–233; the outstandingly detailed material in Richard Colyer, 'Welsh Cattle Drovers in the Nineteenth Century', *National Library of Wales Journal* XVII/4 (1972); Geoffrey N. Wright, *Roads and Trackways of Wessex* (Ashbourne: MPC, 1988), pp. 126–139; and Eric Jones, 'Livestock Movements and Agistment Tithes', *Southern History* 40 (2018), pp. 116–133.
10. Eric Jones and Patrick Dillon, *Middle Ridgeway and its Environment* (Salisbury: Wessex Books, 2016).
11. Derek Hurst, *Sheep in the Cotswolds: The Medieval Wool Trade* (Stroud: Tempus, 2005), p. 189.
12. Confusion may also arise because 'Welsh' was sometimes used to describe any stranger.
13. This conflates the two meanings of ture and must be an underestimate in view of the several places where the term persists as a name for market-place alleys and minor streets. Natural England subsidised the reconstruction of tures in the Stanway district, where they refer to the long narrow walled lanes along which sheep were moved from high pastures to water at dew ponds lower down. Seventeen tures of this type were identified in 1990. At Cutsdean, where tures were established after the enclosure of 1777, the dewponds were segmented by up to six radiating walls. A similar arrangement occurred on Salisbury Plain, where in his village Thomas Mozley found two men working all day to bucket well water into radiating troughs for sheep to drink (*Reminiscences: Chiefly of Towns, Villages and Schools* [London: Longmans, Green, 1885], p. 304). The provision of water for sheep on the high ground was a serious problem. In the eighteenth century the topographical artist, S. H. Grimm, sketched a 'sheep well' like a padlocked sentry box on the Marlborough Downs but how common such arrangements were remains unclear.
14. Gwyn Williams, *The Land Remembers: A View of Wales* (London: Futura, 1978), p. 163.
15. Daniel Defoe, *A Tour Through England and Wales* (London: J. M. Dent, 1928 edition), I, p. 289.
16. T. Mozley, *Reminiscences: Chiefly of Towns, Villages and Schools* (London: Longmans, 1885), II, p. 315. I am indebted to Shaun Kenaelly for this reference.

17. A comprehensive illustrated account is given in Nigel Wardell, '"Far-Famed for Sheep and Wool"—A History of East Ilsley's Markets and Fairs', East Ilsley Local History Society, 2006. See also Jones and Dillon, *Middle Ridgeway.*
18. Brian Smith, *The Cotswolds* (London: Alan Sutton, 1992), pp. 171–172.
19. Fairford Flyer no. 16 (Sept. 2020), Fairford History Society.
20. The continued distortion of Fairford's road system makes the difficulties evident to this day. For road capture, see Eric L. Jones, *Landed Estates and Rural Inequality in English History* (Basingstoke: Palgrave Macmillan, 2018), Chapter 5.
21. This is not to say that, despite its later significance, it was in its entirety a connected Prehistoric track, much less that it is 'the oldest road in Europe'—could such a boast be demonstrated.
22. Rev. Canon Jackson, 'Swindon, and Its Neighbourhood', *Wiltshire Archaeological Magazine* VII/xix (1861), p. 126.

CHAPTER 9

Colonising the Hill Country

Middle Ridgeway

The Ridgeway ran without interruption along the northern edge of the Wessex chalk lands, overlooking the Vale of White Horse. Its stereotypical stage, the 'Middle Ridgeway' between Avebury and Streatley, passed almost no houses, offered a single hostelry (appropriately called the Shepherd's Rest), and for the distance involved was crossed by unusually few north-south roads.[1] This left the belt of high ground along the Ridgeway as an uninhabited 'island', large by Southern English standards, with Iron Age earthworks and Roman sites but little sign of further occupation until recent centuries.[2] The Saxons had built their villages on the spring-line at the foot of the escarpment but they and their successors reserved the downs above for grazing sheep. Drovers found the lack of settlement on the high ground convenient for moving animals unhindered, although they no doubt found it annoying to have to descend the escarpment to find a pub.

Drovers favoured empty stretches where they paid no tolls and their animals were not liable to stray into standing corn. The Ridgeway met their requirements well but the situation began to alter during the seventeenth century, when the demand for food in London inspired more trade in livestock and simultaneously persuaded farming communities to extend the area under crops. The plough moved away from the home fields

E. L. Jones, *Landscape History and Rural Society in Southern England*, Palgrave Studies in Economic History,
https://doi.org/10.1007/978-3-030-68616-1_9

around villages, started to ascend higher slopes and bit into the downland grass. This created a problem for the farmer: he could not lead his plough team further and further from the farmstead every morning and home every night because the horses would consume too much of what they had helped to grow.

Colonising the Downs

The solution was to move the base of operations away from the villages, first by establishing barns and feed-yards well out in the new fields, next by building isolated cottages or pairs of cottages where labourers could live and tend to farm animals, and finally by shifting whole farms up the hill. More exactly, the upland farms were placed in the little valleys that penetrated the downs, where a well could be sunk to obtain the large volume of water needed for animals and human inhabitants. Cultivation approached the Ridgeway ever more closely. Eventually formal enclosure took place, with the effect that the Ridgeway, like drove roads elsewhere, became confined to narrow strips with hawthorn hedges. Previously drovers had let their flocks and herds amble forwards amidst wide views over a sea of grass. Now they had to be kept moving, obliged to graze narrower verges a little faster as they travelled, and consequently putting on less weight.

After the Restoration in 1660, a shift in land use occurred across many parts of the country, not merely on the downs near the Ridgeway. Grass uplands were increasingly colonised, cultivated and converted into regular farms, big by English standards and tame compared with the previous bare wind-blown scenery. Maps, of which the best are the original 'drawings' of the Ordnance Survey in the first decades of the nineteenth century, mark the barns that came to be sprinkled over the hill country and also show tracts with the symbol for rough grazing. The Ordnance maps replaced the schematic charts of earlier times. They were unusual because neither earlier nor later maps trouble to mark out-of-the-way barns or depict types of grassland. What the original drawings were capturing was the high tide of the Napoleonic War era, when ploughing pressed more closely on the Ridgeway than it had done since Iron Age or Romano-British times—or than it would do again until the Golden Age of Farming in the middle of the nineteenth century and the renewed ploughing campaign that started about 1940. The Napoleonic wave of cultivation was the culmination of the late seventeenth- and eighteenth-century agricultural improvement that had put grassy hills everywhere under crops. It was much the same (though apparently earlier) on the Cotswolds, which are in view to the north from the Ridgeway, and indeed happened on open sheep downs everywhere. Dorset was another prime instance among many examples.[3]

In *The Ancient and Present State of Glostershire*, Atkyns observed that the Cotswolds were already arable in 1712. The central stretch of what was then the Berkshire Downs, or more precisely the Lambourn Downs, was preserved as natural grass until later than that. It succumbed to the plough when Napoleon threatened, which led to a landscape like that of the Cotswolds, minus the stone walls—but comparably spacious, with outlying farms, huge fields, cereal crops and folded sheep. The reason why the area around Ashdown and White Horse Hill on the Lambourn Downs was kept under grass for a long time was personal, or if you like idiosyncratic. It suited the Cravens at Ashdown House, owners of the biggest estate in Berkshire, to have much of the area left untouched.[4] Generation after generation, the Cravens were obsessed with blood sports, as one of their descendants admitted to me. They kept wide stretches south of White Horse Hill under grass for hare coursing, until soaring prices for grain meant that even they or their tenants surrendered to the temptation of ploughing them up.

Across the high chalk were littered blocks of hard, coarse sandstone, called sarsens, which got in the way of ploughing. Some were as big as an office desk. They came from a fragmented geological layer on top of the chalk. For centuries they had been hauled off the hills to use in house building and road building in the vale, with the effect of freeing higher ground for cultivation. Dateable houses in the district provide some indication of the long periods when at least some clearance took place. Activities took a step upwards between 1850 and the late 1930s when a sarsen cutting firm was in operation at Fyfield near Marlborough.[5] The size of the territory originally dotted with sarsens must have been enormous. In the years about 1960, when machinery was available to do the work, land clearance was resumed with a vengeance. A particularly herculean effort was then made in the area north of Up Lambourn, where remnant sarsens are still in view along the tracks and next to farmhouses, despite many having been presented to towns as far away as Swindon for whatever ornamental purpose they chose.

Knighton Bushes was a prime site for this episode of clearance.[6] As the 1960s began it was a sizeable patch of rough grass dotted with old hawthorns, good for remnants of the area's former grassland birdlife and containing a sarsen field of its own. Reclamation has left only a shrunken remnant of the type of landscape that the great Oxford geologist, W. J. Arkell, termed 'thorn savanna'. It is a tiny memento of the rock-strewn landscape which for millennia sloped southwards from the line of the Ridgeway. Today sarsens are best seen in the dry valley by the road east of Ashdown House—that Amsterdam doll's house built by the first Lord Craven as a hunting lodge for Elizabeth, the Winter Queen. The sarsens here harbour lichen of European significance.

The 1950s and 1960s saw another significant subtraction from the old environment, the pulling down of barns and outlying feed-stalls. Originally denoting cultivation's high tide, they had often been named by colour, Red Barn, Buff Barn and so forth, as a help to finding the way in a countryside that might strike the visitor as featureless. In times of depression, building standards dropped, plank walls were creosoted in the name of economy and thatched roofs replaced with corrugated iron. Outlying barns no longer found much use and the adoption of machinery after the Second World War rendered them almost redundant. Most have gone now. So many were left empty that an enterprising local doctor arranged for five to be taken down and the material re-used to build a Youth Hostel on the Ridgeway above Wantage. In the 1960s I used to shelter from the rain in some of those that remained and passed the time examining graffiti on the beams. Most of the scribbles were no more than that, scribbles, but occasionally a date could be found from as far back as the eighteenth century, witness to a major phase in the arable colonisation of the downs.[7]

Land-Use Fluctuations

On a timescale of centuries the high downs have more than once switched between short-grass grazing and arable land. This fluctuation requires a little more definition and we can pick out two phases of special interest. The first phase relates to the downs and wolds of Southern England as a whole, as opposed to the intervening clay vales. The light land and heavy land ecosystems were distinguished by the (literal) horsepower needed to plough their soil. The downs, it was said, 'ploughed one horse lighter'. In a broad sense, allowing for time-lags and scattered exceptions, the two ecosystems underwent a major, although gradual, inversion during the period 1650–1750. Fodder crops such as clover, sainfoin and turnips were increasingly adopted in the hill country and made it possible, via manure from the sheep-fold, to keep the light soils 'in heart'. Because they were cheaper to cultivate, these dry grasslands were more and more converted into arable. Against this competition, heavy land farmers in the wet clay vales below put more of their land under pasture and slowly reduced the extent to which they grew grain for the market.

Among the results, the transhumance or seasonal movement of livestock from the high ground to the richer pastures of the vales was curtailed, although it was revived in the driest years when fodder became scarce on the hills. Ordinarily, the 'new' fodder crops made it possible to feed more animals in the uplands and feed them there for longer. The effect of the land-use inversion was however subject to one modification: the populations of towns and villages in areas of only one soil type tended to stall, decline or actually shrink but those situated along the junction between hill and vale (Wantage for instance) tended to prosper. They could facilitate trade between the ecosystems. After the areal transition the chalk hills had grain to sell, besides mutton from the flocks folded on their arable fields, whereas the clay vales turned to specialising in beef and cheese. Locations that brokered exchanges of products could balance gains in the prices of one type against downturns in receipts from the other in both the short and long run. A similar buffer against adverse fluctuations was mirrored, fractal style, within those individual farm businesses which straddled hill and vale.

The second phase included the depressions that affected agriculture after the Napoleonic wars and again from about 1880 until 1940—the latter relieved only by a brief boom lasting from late in the First World War until 1922. In the runs of depression years, arable farming retreated. The start of the twentieth century saw a large expanse of Ridgeway Country acquired and turned into a sheep ranch by an entrepreneurial butcher from Ramsbury. He was accused of bringing about the depopulation and eventual desertion of the hamlet of Snap, near Aldbourne. An unexpected emigration took place from Ramsbury and Aldbourne to Patagonia and at least until recently it was possible to meet in these villages the children of some of those who went—returnees who were Argentine-born. There was no adequate or lasting agricultural recovery until the Second World War and the very title of Sir William Beach Thomas's book, partly based on the area around Aldbourne, tells it like it was: *How England Becomes Prairie* (1927).

Little investment was made between the wars. This is not surprising because the Berkshire Downs were harder hit by the depression than most parts of the country, having thin, poor soils with little moisture in them. Between 1921 and 1938 tillage fell by 27 per cent; on the other hand cattle numbers rose by 120 per cent and sheep were up 11 per cent.[8] Local agriculture was accommodating to the steep fall in cereal prices and somewhat lesser fall in the prices of livestock products. The author of a highly numerical study goes to lengths to determine how far scrub spread during this period. He admits rather half-heartedly that scrub did increase but does not make allowance for another index of retrogression: the marked reversion of land to coarse grass, which local memory and contemporary writing describe. His conclusion is that the Berkshire Downs suffered less between the wars than they had in the 1880s depression, yet his evidence on the latter is geographically and chronologically misleading: he states that in the earlier depression the village of Snape (sic) on the Berkshire Downs became derelict.[9] But Snap is in Wiltshire and was not finally deserted until 1914. The last resident, Rachel Fisher, found herself alone in the remains of the hamlet, her son having gone to Patagonia. She was moved as an act of charity down to a cottage in Aldbourne by one of the Cooks, a local family of builders.

Flora and Fauna

When she was evacuated, which is what it amounted to, an era seemed to come to an end for the high ground bordering the Ridgeway. Plough agriculture had given way to sheep ranching. Twenty years later a topographer could describe the vicinity as having 'sheep adrift in hedged acres where thistles abound, while one shepherd rides across the land which the life of a whole village once kept clean and fruitful'.[10] True, but this was not the first time land use had swung one way rather than the other, nor would it be the last. Druce, the Oxford botanist, had already pointed out in 1926 that 'the flora is continually changing, and never has it been in such a state of flux as at the present time'.[11] He referred to 'relic floras' on the Ilsley Downs escaping the destructive forces of what to him was modern life, and later botanists applied the term to patches of plants surviving in the high chalk grass about White Horse Hill.[12] What has changed can change again and alter the whole aspect of the landscape. One novelty today, rarely remarked on, is the emergence of long stretches of tall hedgerows and trees along the Ridgeway, attracting species not present during the long years when the high downs were bare. It affects especially butterflies.

At any one period the main elements of wildlife in the Ridgeway country were correlated with the dominant land use, give or take slight lags. William Hewett Junior's history of Compton Hundred, east of the White Horse and including East Ilsley, caught one of the big alterations on its cusp at the start of the 1840s.[13] Sheep walks, he wrote, had gradually yielded to the plough and were under luxuriant crops of cereals but there were still extensive downs that fed lots of sheep. Partridges, pheasants, land-rails (corncrakes), and quail abounded in the open fields—he probably meant fields which were physically open rather than in communal ownership—while dotterel and golden plover 'amuse the sportsman'. Corncrakes and the migratory 'trips' of dotterel are no longer found, while a species that Hewett does not even mention had already vanished. This was the great bustard, which would be called in modern cant the 'iconic' species of the chalk lands. Being the size of a turkey it would not have been difficult to shoot but had been preserved for the sport it offered to hunters with dogs. The great bustard was last reported hereabouts in 1802. That coincided with Napoleonic-era ploughing and a reasonable surmise is that the species fell victim to land-use change, as it did elsewhere in Wessex.[14] Nevertheless, what amounted to commercialised hunting cannot have helped: Hewett remarked that gentlemen with packs of harriers used to visit the Ilsley Downs from Surrey, Buckinghamshire and Gloucestershire. Presumably they were coming to hunt over grass downs soon to go under the plough in the giant High Farming boom of the 1850s and 1860s. The interest of the downs for the naturalist was about to shrink.

Notes

1. Eric Jones and Patrick Dillon, *Middle Ridgeway and Its Environment* (Salisbury: Wessex Books, 2016).
2. Plate V in W. J. Arkell, *The Geology of Oxford* (Oxford: Clarendon Press, 1947) depicts the vicinity of White Horse Hill as a grassy 'island' amidst what by that date had become a zone of cultivation.
3. J. H. Bettey, *Discover Dorset: Farming* (Wimborne: The Dovecote Press, 2000), p. 36.
4. By 1870 the estate of the Jones-Loyds of Ardington had eclipsed that of the Cravens in extent, but only just.

5. N. E. King, 'The Kennet Valley Sarsen Industry', *Wiltshire Archaeoogical and Natural History Magazine* 63 (1968), pp. 83–93.
6. The best-known clearance had taken place at Avebury, where in the seventeenth century local farmers learned how to break up the sarsens and did not hesitate to apply their destructive method to some of the stones of the Avebury Circle.
7. Historic barns survived later on the Yorkshire Wolds, another area of chalk upland, and have yielded more graffiti. They had been constructed to permit the winter feeding of cattle and obtain the dung to raise the fertility of thin soils. See Colin Hayfield, 'Vessey Park; The Development of a Yorkshire Wold Farmstead', *Yorkshire Archaeological Journal* 70 (1998), pp. 109ff., and Kate Gils and Mel Gils, 'Signs of the Times: Nineteenth-Twentieth Century Graffiti in the Barns of the Yorkshire Wolds', White Rose e-print (2010).
8. R. B. Tranter, 'Agricultural Adjustment on the Berkshire Downs During the Recession of 1921–38', *Agricultural History Review* 60/2 (2012), pp. 214–240.
9. Tranter, 'Agricultural Adjustment', p. 238.
10. H. W. Timperley, *Ridge Way Country* (London: J. M. Dent, 1935), p. 183.
11. G. Claridge Druce, in James J. Walker (ed.), *The Natural History of the Oxford District* (London: Oxford University Press, 1926), p. 121.
12. For other nineteenth- and twentieth century naturalists in the district, see Jones and Dillon, *Middle Ridgeway*, pp. 110–123.
13. William Hewett Jun, *The History and Antiquities of the Hundred of Compton, Berks* (Reading: John Snare, 1844).
14. A report of a large flock hunted with dogs on the Hampshire chalk in the 1820s is exceptionally late. Dorothy Beresford, *Nether Wallop in Hampshire* (Over Wallop: Privately published, revised edition, 1989), p. 30.

CHAPTER 10

Parkland

Fairford Park and Landscape

Fairford is a small place in a corner of Gloucestershire, close to the head of navigation on the river Thames, where the county meets Wiltshire, Oxford and until recent times, Berkshire.[1] From my study window I overlook an overgrown ha-ha or sunk fence on the edge of Fairford Park. The park landscape, legally intact and held under a trust deed, preserves the residue of what was an elaborate and highly fashionable property. Here, at the Restoration, Valentine and Thomas Strong, masons whom Christopher Wren employed on St. Paul's Cathedral, created a mansion for Andrew Barker, the son of a Bristol merchant.[2] The subsequent history of Fairford Park, individual in its detail, did not differ greatly in outline from other country houses and their surrounding estates, although much of it has to be assembled from landscape clues rather than written records.

Fairford had been an outlying estate belonging to the old family of Gloucestershire landowners, the Tracys, who sold out about 1650. It was then bought by Barker, who had inherited a substantial fortune and was well placed to buy up the distressed estates of Royalists like the Tracys who had been wrong-footed by defeat in the Civil Wars and consequently subject to heavy fines. Barker's new mansion was set back in a park and therefore conformed to the objection made by Roman Catholic historians that the Protestant rich drew away from their local communities

E. L. Jones, *Landscape History and Rural Society in Southern England*, Palgrave Studies in Economic History,
https://doi.org/10.1007/978-3-030-68616-1_10

into splendid isolation. Valentine Strong had more usefully devised a type of fireproof malt-house but there is no doubt that the trophy house he and his son put up for Barker was aesthetically impressive.[3] The house was remodeled about 1740 by Barker's granddaughter, Esther Lambe, and her husband, and a little later they also modernised the grounds.

Esther Lambe made her late husband's nephew, John Raymond, her heir. He changed his surname to Raymond Barker and moved to Fairford about 1780. With banking money from his own marriage behind him, he set about further changes to the park landscape in 1783–1787. His landscaper was William Eames, a pupil of Capability Brown, though one who preferred more trees than his mentor and made a specialty of water features. This period, the 1780s, saw the creation of elaborate gardens at Fairford Park and also saw the park pushed out to the east, overrunning the road to Quenington and obliging road traffic to travel further than before. This was perfectly legal: an Act of 1773 had made the privatisation of public routes straightforward, but required the landowner to provide an alternative route if two magistrates found it necessary. From Raymond Barker's point of view it would have been advisable in any case, in order for local business to be carried on.

The estate system locked in this convenient arrangement, which meant simultaneously that the crossing of estate land, and not merely access but the view too, were denied to the public in a fetish for privacy. As was written about Scotland, 'the gentry about Edinburgh had a peculiar idea of property, enclosing it with these high walls, as though they were lunatics or a menagerie'.[4] The tall stone walls of Cotswold parks and houses, of which Fairford has several, were erected to block the view. They are very hard, often impossible, to date. By comparing illustrations, the walls shielding Keble House, Fairford, home to the father of John Keble, author of the best-seller *The Christian Year*, can be assessed as dating to about 1820. Earlier walls are occasionally mentioned in Fairford deeds but the early nineteenth-century seems to have been the peak time for middle-class withdrawal from Fairford's community. The Barkers, like landowners elsewhere, had already withdrawn into their park.

Erecting walls around the homes of minor professionals symbolises the class-structured society that still prevails. Any paradox of the coexistence of walls and ha-has is apparent rather than real. Privacy was ensured for the great house by the size of its park, not to mention a staff of gardeners and gamekeepers; a ha-ha at the edge might let the house be glimpsed but not approached closely, so that a wall was unnecessary. The occupants of smaller properties were less able to retreat from the gaze of hoi-polloi. Their houses were closer to roads and the gardens not large enough for ha-has to shield them from public scrutiny, hence the resort to high walls.

After Andrew Barker's time, Fairford Park passed through another two or three Barker and Raymond Barker generations until in the twentieth century it was let and then sold during the Second World War to Ernest Cook, of the Thomas Cook travel agency family. Cook could afford the hobby of buying landed estates and ensuring their preservation. In the 1950s he offered Fairford to the National Trust, which prevaricated, possibly because it needed endowments to run the flood of properties coming its way in that decade, whereupon Cook became impatient. He established his own trust which now owns and runs a portfolio of estate properties with headquarters in the former estate yard at Fairford.

Barker's great house has gone. It was demolished in the 1950s in the rush by country houseowners to offload unmanageably expensive properties, many of which had been damaged by military occupation during the Second World War. More than one authority has labelled the demolition of Fairford Park the greatest loss of a country house in Gloucestershire. The architectural historian, John Harris, watched it being gutted in 1955, to be replaced on the same site by a secondary school whose architect, he dryly noted, was an expert at building fire and police stations.[5] Opinions may differ as to whether a house for one rich family would have been more valuable to society than a high school. Architectural historians and garden historians rarely consider the opportunity cost of building and retaining historic mansions, stables, kennels, walls, gardens and so forth. Their threnodies were at a pitch in the 1950s when, in what has been called the accursed year of 1955, one big house was being demolished every week. Yet since that time, far from decreasing, many new country houses have been built, with revived anti-social effects in terms of restricted access and blockages of public rights of way.[6]

What was left of the gardens at Fairford Park did not outlast the wartime use of the house and inner deer park, at first for an American forces hospital and then as a hutted encampment for displaced Polish families. Many of the latter have been absorbed in the local community, but the camp has been razed to the ground. The skeleton of the park is what remains, avenues of trees still standing and cattle grazing where the huts once stood. A vista running one mile to the north, beyond the upper deer park, ends where an eighteenth-century obelisk rears up in an arable field. Stone walls line the roadside of the park while the interior compartments are separated by deep ha-ha ditches with some faces lined with stone. Deer could not leap out but would have been in view as a picturesque sight when the ornamental landscape was in full fig. The land is converted to agricultural uses but the boundaries are the old ones and the historical layout is discernible enough.

Historical? It depends on the period. What is visible now is essentially the park preserved in the last stage of the estate's occupation by a single family, minus its centrepiece of Barker's mansion, with the setting not very obtrusively used for modern farming, forestry or building. The general plan of the Georgian landscaping survives, with Victorian touches such as laurels planted for pheasant cover. Plenty of great estates continue more intact than Fairford Park and have kept their mansions and archives, but this estate was rescued only at the last minute and is held in suspended animation. Many of the attractive but essentially frivolous appurtenances of a pleasure park remain. The flower gardens may have gone but the old form of tree planting is to be seen, give or take gaps in the age-structure of specimens in the deer parks, although the oaks are stag-headed and past their prime. Built ornaments are retained, such as the Cascades where water from the River Coln tumbles on its way to the Broadwater, which is a mile-long stretch of the river, dug out and widened to enhance the scene. Eames did some of the work but the feature pre-dates him. There is a boathouse of a distinctly Victorian cast and where the water narrows again and was made to power a corn mill the Broadwater is crossed by an elegant Palladian Bridge.

The park through, or rather alongside, which the Coln runs was intended to form pleasure grounds, dotted with trees but mainly under grass. Grazing by a few cattle or sheep and mowing hay with the scythe would not have constituted maximum usage. Timber may have been felled for firing or building but the trees were often ornamental species of little use to a saw-mill. If it did not quite sterilise the ground, the home park restricted its flexibility. Like other landowners, the Barkers were willing to pay the price for amenity: their primary goal was not profit, which was (presumably) secured from investments and from the rents of tenant farms on an estate that grew in size over the generations.

Opinions differ about the commercial value of parkland. Charles Quest-Ritson is of the opinion that parks were designed to control animal movements and protect the land from overgrazing.[7] They might be dear to build but were cheap to maintain; grazing could be let and pasture was perhaps 50 per cent more valuable than ploughland. As a result parks were, he thinks, net earners. James Chambers, on the other hand, declares outright that parks were 'wasteful of good farm land', which is most credible in those places where the ridge-and-furrow of former ploughing is fossilised under the grass.[8] The most likely conclusion is that parks contributed something by way of income without achieving full use of the land. Lack of documentation means the issue cannot be resolved at Fairford, where a former estate manager ordered workers to burn an unknown quantity of estate records (though I know of two items that were rescued).

Comprehensive evidence is unavailable with respect to the economic basis of the estate as a whole.[9] Only hints emerge about the sources of the Raymond Barkers' wealth. Innumerable casual references may be found in print about the family but reveal less about the income side of the balance sheet than about expenditure. There are repeated references to genealogy, to the interior of the house, and the bric-a-brac (including some rather poor daubs by family members) which from time to time finds its way into the hands of antique and picture dealers. The family's public role was not particularly conspicuous. In the nineteenth century successive heads served, not surprisingly, as magistrates and in various honorific positions. John Raymond Barker was a vice-president of the British Archaeological Society at the time of its Cirencester meeting in 1868 but that was as intellectual a pursuit as comes to light and probably meant no more than stumping up a subscription. Otherwise, once Andrew Barker had secured for Fairford a charter for a weekly market and two annual cattle fairs, the family heads presided over what was essentially a showpiece estate.

Agriculture

Sir John Clapham may have been right when he said that economic history is merely a foundation and exists to carry better things, but it seems perverse to ignore activities which were the *sine qua non* of luxury consumption and lives of leisure. This calculated veil of ignorance means that it is only now, in 2020, that the fact of one thousand estates gaining from slavery, by proxy of colonial sugar production, is receiving attention. At least slavery does not seem to have funded the Raymond Barkers. Details are however scarce about the fundamental basis of their Fairford estate business, farming. Esther Lambe did take the opportunity to buy land at the enclosure of 1769, which set or froze many field boundaries. The 'mounds' which resulted can be traced today—'mound' being a term in Gloucestershire and adjacent counties for any field division, whether bank-and-ditch, stone wall or hedgerow. What the husbandry routines were on those acres at the time of enclosure or afterwards is unknown but occasionally there is the spoor of a vanished practice. The most notable is the term 'milking path' for what is now a footpath through a housing estate. It was where cows were fetched back from the common and is found in one or two other Cotswold locations.

Some of the Barkers followed agricultural fashion. According to Thomas Rudge's *General View* of the agriculture of Gloucestershire, written about 1805, John Raymond Barker had brought in New Leicester Sheep (Bakewell's breed) to improve his flock. Next there was a shift to the Southdown breed and in 1834 his son bought some of those from John Ellman of Glynde, Sussex, the leading supplier. This Barker must have been impressed because he was a Southdown judge in 1840. John Raymond Barker's grandson, George Raymond Barker, was honorary secretary of the Cirencester Agricultural Society from its founding in 1830. In that decade the family was sufficiently involved in agriculture for J. R. Barker to write to the *Farmer's Magazine* in 1839 asking whether anyone could tell him if saltpetre is good as manure.

An unexpected contretemps throws a sidelight on the family's involvement with farming. Thomas Raymond Barker, a younger son in the family, was the defendant in one of the longest legal cases of the nineteenth century. In 1811 he was the co-respondent in a case of 'criminal conversation' resulting from his seduction of Anne, wife of Edward Loveden Loveden (sic) of Buscot Park, a real 'what the butler saw' tale. He thereafter lived with Anne and farmed at Hambleden, Buckinghamshire. When she died in 1821 she left him a great deal of money and he remarried. He was farming 300 acres and employing thirty-six men and boys on his death at 73 in 1851.

The opprobrium one supposes Thomas had earned was no barrier to public recognition, for he became a magistrate in Buckinghamshire and Oxfordshire as well as chairman of the Thames Commissioners, besides serving on the committee of management in 1840 of what was at first called the English Agricultural Society. He and his father had both subscribed to that society in 1838 and he was to chair its finance committee. It might be thought his own legal ordeal would have imbued him with charity but not a bit of it. At the Buckinghamshire Quarter Sessions in 1834 a man was transported to Australia for stealing one of his sheep—transported for life.

After the first thirty years or so of the nineteenth century we enter a period when surprisingly little is known about farming and the Barkers at Fairford. Their titular involvement with agricultural societies leaves almost no other trace. The tithe map of 1841 depicts the fields to the north of West End Gardens as mostly arable whereas the area might have been expected to stay the grassy wold for sheep that it had presumably been for centuries. The field boundaries of the 1830s and 1840s are evident; several of the hedgerows have now grown over small ponds which are marked on the six-inch Ordnance Survey maps of pre-First World War date. The fields were thus suitable for livestock but this may have meant flocks of sheep folded on ploughland rather than grass flocks and it is quite indeterminate how old the ponds are. Conceivably, land that had become arable by the date of the tithe map might have been laid down to grass again for livestock farming during the Great Depression of the late nineteenth century.

The Raymond Barkers may have loomed over the life around them but did not always have it their own way in the face of political dissent. The Tory Radical, political activist, journalist and farmer, William Cobbett, was a visitor to Fairford ('this pretty little town') who used in the 1820s to stay with Richard Iles, apparently John Raymond Barker's tenant at Park Farm. Just off Park Street, in the old environs of the farm, stands a False Acacia or Locust Tree (*Robina pseudoacacia*). I like to fancy it was planted by Cobbett, or through his influence, or is a successor to one he did plant.

Fairford was a stop on Cobbett's *Rural Rides*. He rode from the house of one prosperous sympathiser to the next, such as Budd (a Newbury lawyer) at Burghclere, and Blount, a larger than life farmer at Hurstbourne Tarrant near Andover. Iles took him from Fairford to the Arkells of Kempsford, who were sufficiently impressed to name a son William Cobbett Arkell and the Cobbett Arkell name passed down the generations. And from Kempsford Cobbett went on to Tuckey at Haydon, Swindon. These were independent-minded farmers not all of whom kow-towed to the landowners.

Cobbett loudly touted the benefits of three crops: Indian corn (maize), Swedish Turnips, and the False Acacia or Locust tree. He was right about Swedes, which weather frost better than common turnips, but the maize varieties then available did not like the climate, and Acacias were too inclined to split. Nevertheless he made money from them and sold 1000 trees to Lord Folkestone at Coleshill, where one was hanging on when

I looked a few years ago. A further survivor is to be seen at Eynsham, outside Oxford, next to the house of another of his hosts, Swann, a printer. Perhaps Cobbett gave it to him; perhaps he gave one to Iles too.

Iles was as great a fan of Cobbett as were the Arkells. From 1826, together with J. and S. Vines of Fairford, Richard Iles (sometimes given as Robert, doubtless by mistake) was engaged in raising funds to elect Cobbett to Parliament. What Barker thought of such insubordination on his doorstep may be guessed—he was Captain of the local military, the Fairford and Cirencester Troop of Yeomanry Cavalry, raised in response to the Last Labourers Revolt of 1830. Iles was not deterred. In 1831 he wrote a trenchant letter in the *Political Register* describing his maize crop. And in 1832, with the labouring population as part of the intended market, he placed an advertisement for 'Cobbett's Corn, raised last year by Richard Iles, Fairford, Gloucestershire, On Sale for Seed'.[10] Why did Barker put up with this? A likely explanation is that he was as reluctant as most landowners to eject sitting tenants who had working capital, for fear of having to take farms in hand. That would have meant supplying not merely the fixed capital (buildings, accommodation roads and so forth) which was the landowner's function but the working capital of livestock and implements too, besides having to assume the commercial risks of farming ordinarily shouldered by the tenant farmer.

The Middle Classes

Another devotee of Cobbett was the S. Vines just mentioned. He and his brother Joseph had come from Nonconformist stock at Brinkworth in Wiltshire. They were farmers and maltsters but dissolved their partnership. Later in the century Samuel Vines became an ironmonger and landlord of cottage properties. He was something of a polymath, the first man to take the trouble to preserve finds from the Saxon cemetery discovered at West End, Fairford, besides being a member of the Art Union of London and not a bad painter himself. His wife died at 37 but he lived to be 96. At his death in 1892 he left his property to the two young maids who had looked after him. The property included his 32-room house, Vines Villa, where he grew vines (what else?) against the garden wall—he put up no front wall against the public gaze. He added on the roof a wooden observatory, reached from indoors by stairway and ladder, in order to survey the heavens from windows on each of four sides. The fixtures where it was attached to a chimney stack may still be seen. Some Victorian working-class amateur astronomers did good work, which is now being studied, although it is hard to trace because they did not publish or network with their betters.

But was Vines working class? No. Neither was he really 'gentry' as trades directories list him nor a 'gentleman' as he appears on deeds. He was a middle-class shopkeeper with intellectual interests, accompanied by enough money and confidence to mix with Fairford's few professionals. Three or four other members of the Art Union of London lived here, at least two being physicians or surgeons, with one of them acting as honorary secretary of the local branch. Minor shop-keepers were more numerous inhabitants, either residing along the through road (now the A417) which on this stretch consists of London and Milton Streets or living above their shops in the Georgian-fronted townhouses on the high street. John Raymond Barker did not have as much influence over this segment of society as might be thought from his wealth and position. He can scarcely have been pleased when Richard Iles hosted Cobbett and advertised his own locally grown Indian corn or maize for sale. Cobbett's opinion of Raymond Barker may be gauged from his listing of Fairford Park in the *Political Register* of 1833 among 'noble seats of princely magnificence'.[11]

Nor could Barker persuade 'the principal inhabitants' to be sworn in as Special Constables at the time of the 1830 riots, which were riotous indeed at Fairford. Why would these men refuse what he wanted? Ordinarily shop-keepers and professionals were as likely as he was to fear damage to their properties. They were unlikely to sympathise with underpaid and down-trodden farm labourers, given the perpetual tendency of those in their class to try to elevate themselves above a stratum of society from which many of them had emerged and to whom they were sometimes related. Yet the labourers were their customers. There is much that is obscure but it comes down to the fact that Fairford had a long-standing Nonconformist element distinct from the Anglican clergy and Tory landowners.

Andrew Barker had put himself at odds with the Independents as early as 1660, when he refused to hear their appeals against the breaking up of their meetings. Nick Kingsley surmises him to have been a 'Closet Royalist', given that he was so quick to erect a new mansion in the Restoration year, and his immediate conflict with the Nonconformists makes one think he was just then coming out of the closet to which the Commonwealth would have confined him.

The Fairford Baptists built themselves a chapel in 1723 and in its grounds are tombs that almost mimic those in the churchyard, including what Pevsner's *Buildings of England* refers to as a 'gargantuan Neoclassical one' of 1819. There was money here and the larger farmers in the neighbourhood must have had money too, at least after seasons of high product prices. In the nineteenth century the less well-off in the town erected smaller chapels, consistent with the customary inference that the labouring classes had become disgusted with the Anglican hierarchy for siding with the landowners during the 1830 riots. Yet allegiance was not always so simple: Samuel Vines buried his young wife in a large tomb quite close to the church door and joined her there when he died some fifty-five years later. One might have expected him to be a leader among the Nonconformists, not that burial in the Anglican churchyard was forbidden to such people.

There can be no doubt that a parallel Fairford existed, one that held its skirts tight against the power of the Raymond Barkers, looming as they did in their park right at the top of the High Street. The Nonconformist activity speaks of bravery and independent-mindedness, the more so because the Barkers continued over the years to buy up more land and house property. Some who 'resisted' must have been, or fallen into being, tenants of the estate whose expansion is to be seen in the characteristic gate pillars about the town. They occur where a route from the path called Love Lane stretched towards the pre-Beeching railway station. They also stand at the entrance to the drive of Morgan Hall, the former house of an old Fairford family. Incidentally, this had in its grounds the largest commercial art studio in the country in Victorian times, to which the great and good of the cultural world came, including John Singer Sargent. The Raymond Barkers had at one time taken it over as their dower house.

Buildings and Investment

There was continuity in the location of the main estate farms. Two big, blocky, farmhouses, Park Farm and Milton Farm, were built during the 1860s. Both farms replaced older buildings and barns, and outbuildings survive at the latter from the time when it was called Bye Farm. Both farms in their present guise were put up during an optimistic decade when grain prices were high, ironically ready for the arable slump at the end of the 1870s. Waiten Hill farmhouse is of the 1890s, but had a predecessor whose footings are in 2020 the site of a stalled archaeological dig, while one of its barns bears a date stone of 1790.

A mile or two further out the pattern of agricultural settlement was similar to that on the chalk downs along the Ridgeway (which is visible as Fairford's southern horizon) or on other lower slopes of the Cotswolds to the north. A barn or feeding stall was established on the outer fields to eliminate the journey-to-work for horse teams, which would otherwise have had to be led too great a distance from the main farmstead. The next stage was to upgrade the barn to become a separate farm. That does not seem to have happened on Barker land, where the colonisation of the more distant fields halted at the halfway stage of establishing barns and in one case a sheep-dip which was in use until the 1970s.

The truly outstanding remains of agricultural investment are the sizeable ditches dug, or deepened, to take water off the fields of what is now Milton Farm. Wet or dry according to season, the ditches can be followed readily enough on the ground. Their dendritic pattern culminates to the south, at Horcott, in the major ditch called the Grand Drain. Despite much searching, little has been found about this large-scale project apart from a fold-out map in Rudge's *General View*. A fuller record might be expected of such a feat of water engineering, undertaken in the 1790s by Edward Webb, surveyor from Stow-on-the-Wold. Webb was the employer of William Smith, the 'father of English Geology', who in 1801–1802 was engaged on a similar project not far away at Down Ampney.

The Grand Drain still runs in most seasons. Until an airfield was built in 1943 and caused it to be put underground, it ran across to Washpool lane in Kempsford and then south and east to debouch into the Coln at Dudgrove. Being 12 feet wide at the lower end and nine feet deep in places, it was no trivial piece of work for men with only pick and shovel. At a couple of points water wheels were used to lift one stream above another, though they have gone now. The investment required will have been considerable but we do not know who provided it or inspired the scheme. Since the aim included draining Barker land at Fairford, it has to be assumed that John Raymond Barker was involved, but unless or until fresh sources appear the matter remains uncertain.

Another sign of old estate farming is the floated meadow called Lower Green, west of the river between the Mill Bridge and Town Bridge. Used for short periods today as a rough field for grazing cattle, its initial function has long been given up and if passers-by notice it they may not grasp the original purpose. It was meant to provide an early bite of grass by letting water from the Coln flow out of a drowner dug down the west side of the meadow and then run eastwards along a series of shallow ditches until returning to the river just above the Town Bridge. In meadows of this type, water was let on for a couple of days at a stretch and taken off for two more, under control by hatches. Slots for the hatches are still to be seen, most easily just above the one-time cattle shed called the Oxpens. The hatches have gone but the slots have concrete surrounds, suggesting that the system had been developed or revived in relatively modern times.

What are most noticeable—what passers-by sometimes comment on—are the ditches and intervening ridges across the meadow, picked out by rushes in the depressions and in season by buttercups on the ridges. In the growing season the pattern is clear once it has been recognised but is low down and easy enough to miss. (It is like the New Forest gorse, inchoate masses of vegetation until the underlying pattern strikes the observer.) Similar ridging of water meadows and the surrounds of hatches half-buried in the grass are found right up the Coln to the former paper mill at Quenington, which was owned by the Barkers at one time. Such meadows have been characteristic of valley bottoms throughout Wessex since the seventeenth century. Here, the one early mention of floated meadows is a lease of 1752 involving the Lambes of Fairford Park and landowners called Hamblet of Quenington.[12] It refers to a weir or floodgate in the Coln between Milham [mead], Quenington, and Knipes Leaze, Fairford, with the right to raise a head of water to flood the latter. There is no further documentation about the floating of meadows anywhere along the Coln. Who built them and at what date remains unknown. An aerial photograph of 1928 shows the ridge-and-furrow pattern at Lower Green, with a handful of sheep grazing, but no personal recollections survive about the end of floating.

The existence of the meadows demonstrates the intensive use of the farmed part of the estate lying on the opposite side of the Coln and the Broadwater from the ornamental parkland. Three interests converged here, those of the farmer; the miller; and the river-keeper who controlled the hatches and was responsible for the fishing. The Coln was famous for trout-fishing and the stream resembles and is often (mis-) called a chalk stream. A water level on the meadows suitable for any one participant was not necessarily what his fellows would prefer but of the processes of compromise and conflict that must have continued year after year we have no information. It demonstrates that functional landscape history cannot always be written in detail; only estates with complete archives permit such an exercise. Fairford is not so blessed. What it offers is surface archaeology, with almost all the apparatus of an eighteenth- and nineteenth-century estate still visible—minus the central mansion—thanks to conservation by the Ernest Cook Trust. The working of the old estate economy has to be deduced from the landscape evidence, supplemented by scraps of written material, some of them acquired or collated in recent years by the Fairford History Society.

Allotments

Beneath the workings of estate agriculture was the poverty of the working inhabitants, into which the history of allotments in the parish provides an insight. We start with the original large-scale 1849 map of Fairford in the Community Centre. It shows unconventional markings along the road for hundreds of yards up Waiten Hill on the west of Fairford, with others on the east towards Lechlade and along the Southrop road. They look like lay-bys, though surely too numerous for the mid-nineteenth century. The Open Spaces Society's *Information Sheet No. C10*, 'Highway Verges', offers an explanation. It states that strips of land open to the public run alongside many highways between the metalled road and the fences of the adjoining fields. These strips are often irregular and sometimes very wide. According to a legal judgement of 1819, people might pass along them when the actual roadway was out of repair. If a landowner excluded the public by fencing right to the road he became responsible for repairing the road. As a result fences were set back and space left for people to pass. This would seem to fit the road up Waiten Hill, for who would willingly incur the cost of repairing the highway from Cirencester through Fairford towards London? It was used hard by waggons carrying cloth from the Stroud valley, besides plenty of other traffic.

Unfortunately the Open Spaces Society's explanation implies a continuous strip of public land, whereas the Fairford 'lay-bys' break into strange shapes and awkward lengths. And why were verges elsewhere around the town not marked the same way? What were apparently lay-bys obviously had another purpose, the clue to which is that on some maps, notably the tithe map, short sections of the verges used for allotments are numbered. A terrier of Fairford property in 1834, the original of which is in the

Gloucester Record Office, lists owners and occupiers for these sections, which are described as gardens. They were in fact allotment gardens. Their number runs into hundreds.

We have to assume that the roadsides had fallen into private hands—most conspicuously those of John Raymond Barker—and were being let as allotments. How long did the practice continue? Verges were used for allotments in the 1950s but perhaps they had been reoccupied when food was scarce during wartime. There were allotments on both sides of the A417 to the east of the town at that date, as well as a chicken run next to the Waiten Hill stretch of the road as late as 1960. In subsequent more prosperous times cultivating the plots, which after all were a slightly inconvenient walk away, dwindled and ceased.

Nationally, the Allotment Movement started about 1780 as a response to the growing poverty of the time. Gloucestershire was a notoriously bad county for the poor. The misery can be seen in a Kempsford Vestry Book of the late eighteenth century, where one heart-rending example concerns a pauper girl who was to be sent round the farmhouses, one month apiece, if no one would volunteer to take her in. Allotments appeared before 1813 in Kempsford, while the Overseers of the Poor had provided fourteen acres at Whelford in 1800. This had parallels or maybe an actual connection with the only slightly earlier emergence of allotments around Tetbury, which are usually credited with being the first in England. Further details about the Whelford and Fairford situation are not forthcoming. Whelford was extraordinarily early. Fairford looks on the face of things to have been slower, although this may be merely an impression given where there are insufficient data.

If labourers dug allotments to help feed their families, the poor rates could be kept down. This appealed to landowners, who were monopolising the land so that by the 1870s England had less equal ownership than anywhere in Europe except *possibly* Austria-Hungary and Romania. Farmers, who were mostly tenants, tended to resist the Allotment Movement, despite benefiting from cheaper labour—until their minds were concentrated by the 'Swing' Riots of 1830, an outbreak of desperation among workers in which Fairford figured large.

Fairford's landownership was highly unequal. In 1777 Fairford Park had been extended over agricultural strips to the east and was enlarged in the 1780s as private ground which the Raymond Barkers were not about to share. They owned hothouses and greenhouses whereas the poor were driven to the margin and obliged to take allotments strung awkwardly along road edges.[13] In the mid-nineteenth century allotments of standard size were established at Fairford and Kempsford but argument surfaced as to whether there were still enough. Local newspapers of the 1880s carried letters debating whether there were sufficient in Fairford, Whelford and Kempsford.[14] One letter called on Mr. Barker to make allotments and smallholdings available at Fairford but the writer did not expect him to do so because he was unsympathetic, 'like many men of his class'.

The Town and Kip's View

The earliest map providing details of Fairford Park is held by Gloucestershire Record Office and assigned to 1690, although Hilary McKee considers that it probably postdates the important oblique print in J. Kip's series of views.[15] McKee's doctoral thesis describes Fairford Park as a magnificent landscape with unusual axiality formed by long lines of trees. Her thesis is concerned with the minutiae of garden design, observes that the original Barker formal gardens were swept away in the mid-eighteenth century, and adds that between 1727 and 1779 the landscape of the park was continuously (continually?) reworked. This establishes that anything now obvious is most likely to date from the 1770s, with a handful of Victorian elaborations. Moreover the author considers that there is 'no site evidence of any worth', apparently referring to the layout before about 1780.[16]

The lack of early site evidence notwithstanding, Kip's view is a remarkable portrayal of the park's layout in 1715. To an eye unversed in historic gardens, what is shown is highly artificial and rectangular. The detail has since been swept away, leaving only the bare bones to lie revealed within the modern park. Nature had been submerged, even tortured in the name of successive garden fashions. More interesting is the way in which Kip's view extends southwards to incorporate part of the town, on which, as we have remarked, the Park so closely abuts. Did he climb trees, scale ladders or sketch an impression, in short how accurate was his depiction of streets and buildings?

An anterior question is whether Fairford did or does qualify as a town. The surprise is that anything more than a village should have persisted here when it is only four miles from Lechlade, which held a better hand of cards through its status as a river port. Other towns in Gloucestershire are further apart, sharing out the functions of markets for village hinterlands. By contrast Fairford and Lechlade are almost on top of one another. But definitions of a town either have little operational value ('bigger than a village, smaller than a city') or describe diagnostic facilities at some arbitrary date (such as population—seldom static; or the presence of banks—which in many places, including Fairford, have now gone). May we instead distinguish Fairford from the villages around because it has a market, originally certified by the charter that Andrew Barker secured, but later closed, revived again, and teetering once more on closure in the early twenty-first century? Alternatively, should we rely on urban appearance, where Fairford's market place and highly regular High Street raise it above village status? It does not have the look of a village.

If Kip's was a faithful depiction, it is puzzling. The general layout of the through road (now the A417) from which the market place and High Street branch off north towards the park entrance shows continuity but, apart from church and mill, individual buildings do not fit. Kip does not show buildings down the east side of the High Street, which for the most part consists today of fairly elegant Georgian or Georgian-fronted houses, some of them having on the ground floor windows for shops which were already closing in the 1960s. My surmise is that they were built as residences but their central position meant they were well placed to be converted into shops and had a retail phase lasting about one hundred and fifty years. In British Listed Buildings they are all given as of the mid-late eighteenth century except the intrusions of the sometime magistrates' court of 1860 and the (former) Lloyd's Bank of 1901. There are no hints of anything earlier except Laverton House, stated to be late seventeenth century but raised and extended in the eighteenth century, and a reference in English Heritage's Images of England to a coffin lid built into the wall between two other houses. The Bull Hotel, Chanting House and Fairford House are late fifteenth- or sixteenth century in part but lie on the west side of the market place.

Taking Kip at face value, the Georgian High Street would seem to be either a complete replacement of older buildings which he omits or a concerted mid-eighteenth century development (by the Barkers?). This conclusion puzzles the town's historians but it is where the evidence leaves us. Kip, of course, was really concerned with the park, where he is said to have depicted the hedge lines and rows of trees correctly. His sponsors and chief readers would have been from the landowner class. He was in any event too early to record the best two or three middle range houses currently in the town, such as Keble House and East End House, which were built by Richard Pace of Lechlade. As his trade card shows, Pace built a number of rectories and houses of this social level in the district but his plain early nineteenth-century architectural style was eclipsed by the craze for Gothic.[17]

Lesser houses in London Street and Milton Street, mostly abutting one another, jostled for trade along the through road, and convey a minor urban rather than rustic feel. Externally they seem to have altered little, and some may have medieval, certainly Tudor, origins. The point is that they are mainly stone-built, thanks to stone close to the surface in many shallow local quarries. Settlements of stone escaped the fires that levelled other places, sometimes repeatedly, and this benefited Fairford, which never suffered a major blaze. The only signs of fire damage are where external stone walls have crumbled in the heat of the ages and been replaced by brick chimneys without setting the house ablaze in the meantime. Capital was not required for wholesale replacement, while the depressed years from 1880 to 1940 slowed changes of the sort that marked what were then more vigorous places. Post-war planning regulations have further dissuaded or denied external alteration.

Sluggishness

Substantial investments by Fairford residents were made not here but elsewhere, especially in building in Cheltenham.[18] Rev. John Keble of Fairford was a creditor of the entrepreneur, Joseph Pitt, to the remarkable tune of £6500 and in 1844 George Barker spent £650 on building a house in Evesham Road, Cheltenham. In Fairford, few notable buildings were erected in that century; those that qualify include the two lunatic asylums, Croft House (a Pace house) and what was until recently Coln House School. Yet Fairford's atypical specialism in asylums (a sort of farm diversification, a member of the Iles farming family having been one of the entrepreneurs) did not completely secure it against the late nineteenth-century agricultural downturn. Other substantial buildings include the big estate farmhouses which in the 1860s replaced earlier examples. Until the 1950s there was also an almost surprising number of smaller farms close in to the settlement, suggesting little competition in producing food for the local market. These farmhouses and barns can often still be identified. Ridge-and-furrow exists quite close in, as it does elsewhere in the district, suggesting that land next to the settlement was in pasture, with the ploughland that would have expunged old ridge-and-furrow lying further out.[19] It is interesting what physical remains of old arrangements may still be found although once again they are very seldom backed up by written records.

Fairford Park overshadowed rather than dominated the town, which was big enough to contain specialist craftsmen and a number of shops. The town was surrounded closely enough by an agricultural landscape. As James Lees-Milne wrote in the final edition of *Burke's Landed Gentry*, 'during the 1920s the physical aspect of the countryside still resembled what the squires, with the help of Capability Brown and Repton, had made it in the eighteenth century, namely one vast luxuriant park…'.[20] So it was at Fairford, but only in the physical respect. Appearances may not have changed much but after about 1880 agricultural hardship was inescapable, ultimately even for the Raymond Barkers. They sold off the more urban part of their Fairford holdings in 1923, significantly the year after the repeal of the Corn Production Act, a move that threatened farm rents.[21] The houses did not reek of prosperity or high landlord investment: five still relied on well water and the sale catalogue warns of potential problems with shared drains and party walls. Seven householders had the option of paying part of their rent in chicken, one being recorded as owing two couple of fat hens.[22] Cash seemed scarce enough for Fairford to be home to a barter economy! The estate itself was burdened by a charge in favour of a family member. Title on the houses for sale would not necessarily be given, because of what the auctioneers referred to as the wealth of the estate, which in the circumstances sounds ironic.

The town had once been reasonably up to date, having for instance installed gas lighting in 1852 after a public meeting chaired by John Raymond Barker. By the end of Victoria's reign it was scarcely well equipped. In the early twentieth century the streets were in poor condition; numerous houses and the church still display the boot scrapers that speak of mud and horse manure. The author of *Highways and Byways in Oxford and the Cotswolds* (1905) declared circumspectly that Fairford was 'not the thriving place it was when its splendid church was built'.[23] He shied away from saying the place was decaying but called it slow and referred very pointedly to the downward spiral affecting other Cotswold towns. Only in the 1930s were the streets given a light skim of tarmac and some of the outer ones made wider. Decline in its retail functions may have been delayed but arrived in force during the 1960s, with the start of families driving to supermarkets and the consequent closure of local shops. The Raymond Barkers were withdrawing; between the wars they had already let the big house. The tenants who came might be generous (in 1936 one of the Palmers of the Huntley and Palmer family gave a village hall), but cannot have been expected to preside over the town as decisively as established proprietors.

Structurally Fairford has altered less than many places. The new residential 'estates' are on the outskirts. Two types of older occupation are still apparent, the consumption landscape of the park (conspicuous consumption) and the production landscape of farmland and town. They overlapped to the extent that the park was used for grazing and the farmland for hunting and shooting. Both categories have seen change but preserve the lineaments of their eighteenth- and nineteenth-century purposes, maybe better than settlements that experienced greater nineteenth-century growth. The park is not a buried landscape to be excavated by archaeologists but a jigsaw puzzle whose picture can still be discerned above ground. Nevertheless many pieces are missing and the purpose of this chapter is to illustrate what may and may not be possible by way of recovering the past.

Notes

1. English geography is seldom more ambiguous than here: the nearby part of Berkshire was ceded to Oxfordshire in the local government reorganisation of 1974, the head of Thames navigation may at some period have been higher up, while Fairford Park may refer either to the big house (now demolished) or its surrounding parkland, which survives. I use 'place' to defer the question of whether Fairford is a town rather than a village.
2. The most comprehensive account of Fairford Park (house) is in Nick Kingsley's Internet blog on Landed families of Britain and Ireland.
3. Robert Plot, *The Natural History of Oxfordshire* (Oxford, second edn, 1705), p. 257.
4. Andrew Young, *A Prospect of Flowers* (Harmondsworth, Middlesex: Penguin, 1986), pp. 10–11.
5. John Harris, *No Voice from the Hall* (London: Murray, 1998), p. 137.
6. A discussion appears in Nick Hayes, *The Book of Trespass* (London: Bloomsbury, 2020).
7. Charles Quest-Ritson, *The English Garden: A Social History* (London: Viking, 2001), pp. 139–140.
8. James Chambers, *The English House* (New York: W. W. Norton, 1985), p. 114.
9. Gloucester Record Office holds many deeds from a deposit made by Mullings, solicitors of Cirencester, but actual accounts are not listed.

10. I first put my findings about Cobbett, and about allotments (see below), on the newssheet of the Fairford History Society, the *Fairford Flyer.*
11. *Cobbett's Political Register*, Volume 79 (1833).
12. Gloucester Record Office, D1728, Long Leases of properties in Fairford.
13. *Monmouthshire Merlin*, 16 July 1831.
14. *Wiltshire and Gloucestershire Standard*, 25 May 1886 and 6 January 1888.
15. Hilary A. F. McKee, *The Bird's Eye Views of L. Knyff and J. Kip*, Ph.D. thesis, Oxford Brookes University, October 2004, Internet version. For Fairford see pp. 162–165.
16. McKee, *Bird's Eye*, p. 153.
17. Pace's trade card is reproduced in the Victoria County History (British History online).
18. Steven Blake, *Pittville 1824-1860* (online edition, 2018).
19. An outer field on Milton Farm is called Iris Land, which may be a corruption of High Ridge Land. Ridge-and-furrow is said to have existed here within the last fifty years.
20. Quoted by Michael Bloch, *James Lees-Milne: The Life* (London: John Murray, 2010), p. 96.
21. Sale catalogue seen by courtesy of Sydney Flatman.
22. 'The Lifehold Estates of John Raymond Barker, Esquire, 1768-1884', document in the Fairford History Society Archive. It is indicative of the general scarcity of documentation that this had come into the possession of a former employee of the estate.
23. Herbert A. Evans, *Highways and Byways in Oxford and the Cotswolds* (London: Macmillan, 1905), p. 306.

CHAPTER 11

Resources: Fodder

Minor Products

Historically land has been central to life because production and pleasure depended on it. By modern standards yields per acre were very low for centuries, which made land seem scarce despite far smaller populations. Yet before the end of the Middle Ages, when outbreaks of the plague had not wholly ceased, commons were being encroached upon and enclosures and assarts (clearings grubbed up in woodlands) were creeping forward in the vales of the English lowlands.[1] Competition for land became sharper still during the population growth of Elizabeth I's reign. At that time regional society was still governed by unstable mixtures of personality, violence and law, with ample scope for conflict. There were no banks. When money was not, so to speak, held under the bed, it was entrusted to friends or neighbours, which gave pretexts for chicanery and misunderstanding. Among occasions for fisticuffs or lawsuits, conflicts over land and over fisheries in the little rivers and ponds figured large. A dispute about the grazing of tiny areas at Hungerford, Berkshire, was taken right up to a central court in London.[2]

In such a land-based society an extraordinary range of natural resources was long used—a quaint example from as late as Victorian times being how the prickles were shaved off brambles so that the pliable stems could be used for crinoline hoops. The gleanings of the countryside had always

E. L. Jones, *Landscape History and Rural Society in Southern England*, Palgrave Studies in Economic History,
https://doi.org/10.1007/978-3-030-68616-1_11

been gathered but the potential was most fully grasped and the landscape was interpreted in greatest detail by people who lived closest to the margin of subsistence.[3] The Saxons had employed distinct terms for hills and other features which we see no need to differentiate; perceptiveness about the natural world mattered then and continued to do so almost to the present among the rural population. Walls were painted white to hold heat in gardens and butter markets were deliberately held in shady streets, as in Alcester, Warwickshire. Countrymen tuned their skills to minute differences in topography, plant life, seasons and the weather, and made endless adaptations to soil types and microclimates. Examples can be found everywhere in books about cottage life; the small sample in this chapter from an enormous, miscellaneous, array is representative of them. Nowadays, in place of the efforts whereby smaller, poorer and more isolated communities scraped a living, we prefer to consume leisure. Our richer society buys the products it desires, including ones that might well be gathered free from the countryside, such as elderflowers for making cordial.

An old Wiltshire farmer commented to me about his forebears, who had lived for centuries on the same farm and in the same substantial farmhouse, saying, 'they lived on what we waste'. This sums up past reliance on exploiting the land, although admittedly it does ignore the opportunity cost of collecting and processing a multitude of natural products. Nevertheless the countryside's minor products made existence more agreeable when there were few alternatives or shop goods could not be afforded. Many natural items remained essential, particularly for the poor, and the labour of collecting them could sometimes be made a game or turned into a ritual. For rural society as a whole, which until the nineteenth century meant a majority of the population, the agricultural basis of life was of course even more important than all the lesser products together. Concentrating on the crops of wheat, barley or oats, whose importance is reflected in the emphasis understandably placed on them by the literature of agricultural history, does however minimise the profusion found in the economic botany of the past. The number of different cereals found in ancient thatch by the newish science of bioarchaeology may indicate that variety provided a form of insurance for people who grew their own food and were close to the subsistence margin: if the remainder of the crops failed in a given year, at least one might be expected to fare well.[4]

Animal Feed

Before the introduction of root crops, grass or temporary clover leys were the feed of choice. Grassland can be divided on the one hand into meadows which can be grazed after the hay is cut and on the other hand pasture where stock can be put in whenever it is dry enough for them not to poach the ground, i.e. churn it into the mud. The value of the grass crop, and the range of 'artificial' fodder crops that gradually supplemented grass, are hard to underestimate. Nevertheless, compared with cereals, livestock products (except for wool) are underplayed in general histories, despite the attractiveness of meat to the consumer and the role of animals in producing manure for the cereal crops.[5] The importance of the latter point may not be obvious if viewed only through the lens of modern commercial farming.

Over the centuries, and almost to the present for farmers in difficult environments, woody plants made shift as animal feed when the grazing was poor and stocks of hay had run out. In heathland environments gorse was grown and mashed up, while holly was fed in the Southern Pennines and elsewhere, including the New Forest.[6] So was ivy. The use of holly is 'well recorded', says Martin Spray, 'but not well remembered'. Modern farmers may be incredulous but in some areas, according to Jeffrey Radley, holly was formerly 'of great importance before hay and turnip winter feed'. This suggests it had eventually been outcompeted by the spread of root crops, although clover and 'artificial' grasses were more likely to have replaced it first. The decline of holly as a winter feed is harder to trace than the fact of its historical use and it may be supposed to have withered away unnoticed, lingering in only the harshest districts after regular use had ceased elsewhere. Such inferior types of feed were mainly resorted to when needs must; they are vestigial now.

Medieval communities ranked the acquisition and management of wet meadow high among their priorities. For that reason many parishes acquired, or had rights over, parcels of meadow and other land in places some distance away. It was not until 1882 that detached portions were fused with the surrounding parishes in order to simplify the collection of poor rates; portions marooned within a single parish were done away with in 1932; and the last remnants, the boundaries of some ecclesiastical parishes, were not rationalised until 1985. All sorts of ecological complementarities were facilitated by the fragmentation of these places and many expedients were adopted to make them work. An eyewitness in 1920s' Buckinghamshire described the traffic when five villages in the Eton Wick area sent their cattle and goats up to Farnham Common and Burnham Beeches in summer via green lanes, to graze and browse woodlands, thereby releasing the home villages' commons and greens to grow hay for winter feed.[7]

Where parishes actually owned detached portions of land they were likely to contain meadowland, which it may be assumed was the purpose of acquiring them in the first place. Some were miles away. As many as five Oxfordshire Cotswold parishes shared meadow rights at Burroway by the Thames in Clanfield. The potential for conflict is obvious and disputes are what provide the evidence about the frequent but obscure practice of intercommoning. Other parishes had rights over much nearer patches, their boundaries snaking out through their neighbours' land, enabling them to drive stock to and fro. The Saxons called one route at Wootton, Oxfordshire, Hig Weg or Hay Way.[8] Yet in many places the peculiar arrangements which communities had made were starting to disintegrate earlier than the twentieth-century Buckinghamshire instance might suggest; that was a late hangover. Numerous documents record the fact that parcels of common were privatised in a patchy and irregular fashion. Grazing rights over common land, once informally assumed, asserted or acquired, might then be sold on and gradually become commoditised.

Today it is scarcely possible to tell on the ground how important lowland meadow once was. Only scraps are left, conserved by chance and held onto because of their botanical interest. That the habitat was formerly widespread can be surmised from the incessant assertion in the conservation literature that 97 per cent has been lost since 1939 (sometimes 98 per cent and sometimes since 1945). The figures are extraordinarily precise and I supposed had some solid foundation. Instead I was passed from author to author back to a supposed originator, who was candid enough to admit that he too came up with the 97 per cent figure only because it was already so widely agreed! Since then I have however been directed to an immensely thorough analysis that justifies the figure, but which none of the many other references consulted had mentioned.[9] Eloquent descriptions of wildflower meadows in the past similarly show just how enormous the loss has been.

A telling clue to the previous value of grassland is the sheer number of ploughing restrictions in leases. Landowners were anxious to prevent tenants from ploughing and sowing grain at the end of their leases, thereby cashing in the fertility of permanent pasture.[10] The value attached to pasture is apparent to anyone who reads widely in local sources. Great significance is attached to roadside grazing. Rent could be charged for grazing in churchyards. At Goosey in the Vale of White Horse, an agreement was made in 1862 to reassign the churchyard herbage for a term of 486 years![11]

Local protectionism was likely to rear its head when food supplies and the means of production, including the hay crop, were threatened by outsiders. It was an understandable, if ultimately stultifying, reaction to the stringencies of the past. The tiny market areas for produce were revealed when at the court baron of 1622 in Marston Meysey, Wiltshire, complaints were laid that cattle, wheat, barley and hay had been carried off to the next parish, Kempsford, Gloucestershire, only a couple of miles away, and consumed there.[12] At the following court it was ordered that no one was to remove these commodities from Marston—trade was banned and they were to be consumed where they were produced.

The importance of anything that might raise soil productivity led to other expedients that would now seem strange; for instance towns advertised the sweepings of dung from the roads and farmers bid for them. A substantial trade developed in off-farm fertiliser, such as the purchase of London soot by farmers in deep country.[13] The modern mind would be inclined to think waste commodities like soot might not be worth the trouble and expense of carriage but that would be to underestimate their usefulness before artificial fertilisers became available.[14] Our ancestors were so acutely aware of fertility considerations and the different qualities of land that measurements of area were not fixed but made dependent on the expected productivity. A yardland was in principle the area that could support a family. Andro Linklater cites a yardland of good arable land at Stratford bought by William Shakespeare which comprised 25 acres. By contrast, a yardland of rough pasture might have measured over 40 acres.[15] The pitfalls for anyone making calculations about these matters are all too clear.

Carolina Lane makes interesting generalisations about the improvement of grassland as a whole during the sixteenth and seventeenth centuries.[16] Until that period, she says, pasture was not cultivated and contained many species of plant; large tracts of rough herbage were needed to feed livestock. From then on, productivity was raised by the introduction of more nutritious species, especially clover (a legume that fixes its own nitrogen), and by managing the herbage through restricting the number of species sown. Farmers could manipulate the constituents of their grassland, start to use leys of sown grass, feed more animals on smaller areas and detach their husbandry from its abject reliance on what nature had handed them. Instead of being scattered over wide areas of common or heath, sheep and cattle could be concentrated in fields where their dung could have full effect. A little example, illustrative of the partial shift of animal husbandry from its age-old dependence on unmanaged grass, was supplementing with grain the diet of those traditional grazers, geese.[17]

A Clover Revolution

The grandest claims of all have been made for transformation by clover. Clover is credited with overcoming nitrogen deficiency, raising farm output and feeding the growing urban population. A readership steeped in the history of a coal-based industrial revolution is told that, 'clover was the agricultural equivalent of coal'.[18] Just as coal transformed the industrial landscape, clover is credited with altering the look of the countryside: 'the romantic 19th-century landscape was a clover landscape'. In the twentieth century, when coal began to be pushed aside by other sources of energy, clover was likewise relegated by the production of artificial nitrogenous fertilisers. We are left to conjure up in our imaginations the broad expanses of clover flowers alive with honey bees, whose buzzing added to a sound-scape most of us have never known.

Claims for a distinct transformation have been advanced time and again. To cite one earlier author, they were made in 1913 in a concise and astute paper by V. G. Simkhovitch.[19] He argued that only the fodder revolution made it possible to keep enough livestock, not simply to maintain soil fertility but to improve it, and that this was what enabled grain production to be sustained on light soils. Before that the proportion of pasture largely determined how many animals could be kept (the limiting factor being how much hay there was for overwintering) and how much manure could be produced. Since most lands had to be devoted directly to growing cereals for human food, and the value of natural meadows altered from year to year as rainfall varied or streams wandered in their courses, so control over pasture was real agrarian power. Within any given community the survival of the poor and imperative of equity meant it was desirable to keep reallotting the precious scraps of land.

Imitation may be the sincerest form of flattery but trying to prevent something is an even more convincing testimony. Attempts to halt the fodder revolution reveal how much it threatened long-standing agricultural distributions. William Ellis of Little Gaddesden, Hertfordshire, heard that about the 1680s the men of the Vale of Aylesbury, 'strove with great might to suppress the sowing of the [clover] in the Chilterns or hilly Country'.[20] Ellis called clover, 'the Chiltern farmers' friend, and the Vale farmers' enemy'. In 1687 the Hampshire author, John Worlidge, testified that the success of the new crops had been, 'the occasion of the many endeavours that were used by some Northern Graziers to obtain a law to suppress the Improvements in the Southern Parts, lest Grass and Grazing Grounds should become as plentiful in these as in other parts'.[21] Similar reports came from Dorset and Gloucestershire; the prices of permanent pasture in the vales and of floated water meadow along the streams fell as a result of more abundant forage on the downs and wolds.

These changes went a long way towards altering the relationship between light land hills and heavy land vales, as was mentioned in the chapter on Colonising the Hill Country. The hills were no longer obliged to remain the haunt of grass-fed sheep and the vales ceased to be the country's main granaries. The two regions slowly switched roles and both developed more complex systems of mixed farming. A similar alteration was reported about 1726 by the steward to a Lincolnshire estate, who said that farmers on the Wolds no longer needed to rent marshland for feeding their cattle and sheep but could feed them on turnips on their own farms and sell beasts into the low country.[22] Changes like this formed a kaleidoscope, with different fodder crops upsetting husbandry arrangements in different places at different dates, but the overall thrust was clear. The shift in the relative values of different types of land was no less potent for taking place at a slow pace.

Among the consequences was the effect on a mostly earlier innovation, the construction of floated water meadows. They did not vanish; the boost that their early bite of grass gave was still well worth having until cheap food was imported on a large scale during the Great Depression at the end of the nineteenth century. Yet their influence and value per acre were being eroded. Thomas Mozley, for instance, noted in 1880 a great alteration in agriculture on Salisbury Plain since the 1830s and 1840s, with roots (turnips and swedes) supplying the place of water meadows.[23] So powerful was the effect that it had already been noted in the American colonies and the early United States. By 1800, about a century later than in England, the spread of clover and cultivated grasses there was supplying enough hay to reduce the value of water meadow.[24]

Hay Meadows

Hay meadows, meaning river meadows that might flood but were not artificially floated, were a specific subsystem within agriculture. They differed from most vale pasture on the acidic clays because they were neutral grassland, rich in flowering species. Archaeological enquiry shows that the Thames-side hay meadows originated in Roman times.[25] The evidence comes from analyses of finds of various species of beetles and snails, macroscopic plant remains, and horizons of silt which had been washed down the river. By analogy with modern distributions of plant and insect species it is possible to deduce whether, and to some extent when, the land had been used to produce hay as opposed to being merely grazed. In a few lucky cases dateable human artefacts have been discovered, most excitingly a two metres-long scythe from the fourth-century A.D. at Farmoor, outside Oxford.

Flood meadows were intrinsically valuable: haymaking did strip out their soil phosphorus but this was replenished by river downwash. Their natural richness might lead potential enclosers to covet the meadows, but by the same token was likely to increase the determination of commoners to cling to what they had. This may help to explain the late survival of some lot meads. The means of raising productivity in the meadows was however limited compared with freshly enclosed arable on which up-to-date rotations could be introduced. Certainly, private holdings of meadowland could be fenced against other people's animals but that was no great advantage where protecting crops was not an issue. On common meadows, regulations would be sufficient to exclude any unwanted stock, such as bulls of inferior potential, and to keep livestock off until the hay had been carried. A more important feature of the enclosed meadow was the farmer's ability to mow his hay at what promised to be the best time in the season, rather than having to wait for a date prearranged by the majority in a communal system. Inserting pasture closes into the common fields was a frequent move that may even have been as evident in the past, though more difficult to trace in the documents, as the intrusion of plots of clover, sainfoin, turnips and the like.[26]

Keeping drainage ditches clear and avoiding summer floods was vital. Manorial courts were forever fining people for neglecting their responsibilities. But although the distinction between routine defensiveness and actual innovation is partly semantic, it is not entirely so; management should be thought of as attempts at avoiding loss as much as raising productivity directly, either approach swelling the farmer's pocket. Rolling the land and mole-catching may also have been helpful where a good swing for the scythe was required but, although farmers in some places subscribed to mole clubs and estate archives record paying the mole-catcher, the secondary literature barely mentions these practices. Where they were adopted they too would have been means of loss-avoidance.

Drainage schemes and flood embankments were directly productivity-raising. But small projects might merely pass a flood onto one's down-stream neighbours and organising the works over wide areas was tricky. It was done on a large scale at Kempsford, Gloucestershire, right at the start of the nineteenth century. Similarly it was carried out in the 1860s between Northmoor and Eynsham, Oxfordshire, although the aim there was to drain land for arable crops.[27] A big effort at coordination was made. Sometime before the 1860s the largest tenant farmer had organised a scheme among his fellow tenants. It was so successful in drying out the fields that the farmers wanted a bigger embankment and canvassed their landowner, who agreed to become the principal promoter.

The bulk of the documentation about managing hay meadows dates from the last one hundred and fifty years—their 'twilight era'—and should be jobbed back earlier with caution. Altering the mix of meadow plants was difficult; the little available information tends to come from the literature promoting improvements and may not reflect the ordinary practice of early centuries.[28] In the mid-eighteenth century, William Ellis from Little Gaddesden, Hertfordshire, listed nine species as desirable in sheep walks but a visitor, the Swedish botanist Peter Kalm, made two careful analyses of haystacks and found them to contain 21 and 26 species, respectively.[29] Possibly farmers liked the butter from herds fed on rich mixtures of herbage, or they imputed value to meadow herbage because floristic profusion seemed to signal prosperity. Thistles and wild garlic were extirpated, the latter being thought to taint the milk. It is not known whether certain other species were positively preferred and, if so, what could have done to encourage them. Farmers may have been intuiting the modern conclusion that there are benefits from herbs in terms of amino acids and so forth.

Known fluctuations in alluvial deposition seem to indicate that the use of hay meadows collapsed at the end of Roman times, to be resumed under the Saxons. Once the trade in black cattle from Wales and from Hereford October Fair grew after the late seventeenth-century demands placed on meadowland may have risen. The animals were fed along the Upper Thames, breaking their journey there en route to fattening grounds closer to markets in the London area. By the early eighteenth century, black cattle may have been depastured at North Meadow, Cricklade, where the haying strips were reorganised at some stage. Clattinger Farm, five miles away, contained accommodation land for animals in transit, though details are lacking.

Surprisingly, given all that is written about shortages of fodder in early times, in the late eighteenth century the numbers of Welsh and local cattle combined were sometimes too few to eat all the spring and aftermath grass.[30] The hay meadows, or more precisely the production of hay, was an unstable sector within agriculture and created a saw-tooth graph of profitability. From time to time hay was superabundant and its price fell. In other seasons shortages of both hay and grazing could be acute. Because the weather changes so quickly, the market for hay is chancy. On innumerable occasions the hay was spoiled, mainly by floods that literally floated haycocks down one valley or another. Or drought was the problem. The hay harvest was so bad in 1860 that farmers offered to buy the grass growing in the streets of Hereford.[31]

Grass as hay was indispensable to the livestock sector, besides, as mentioned above, being indirectly important to the arable crops via the manure it helped to produce. Nevertheless, in the very long term its value subsided. The point was made by the Oxford historical geographer, Frank Emery, when he observed the widespread opposition in the early seventeenth century to enclosures within the common fields on which individuals wished to pasture their own beasts.[32] Land where the whole community's animals might graze was thought too precious to be abridged. But later that century the greater availability of feed from sainfoin, ryegrass and clover somewhat allayed the fears.

Intensified Management

The history of fodder supplies from the seventeenth to the nineteenth centuries shows a reduction in the relative but not the absolute role of the grass crop. Clover, sainfoin and the like became embedded in rotations and were successively supplemented by root crops such as turnips, swedes and mangol-wurzels. Meanwhile, at least by the nineteenth century, the production, management and marketing of hay itself became more specialised. Hay, once too bulky to be readily carried overland, increasingly entered the market as a saleable commodity with its own attendants, so to speak—members of one of the easily neglected sub-professions within agriculture, in this case hay tiers or hay trussers (travelling reapers, contract shepherds and teams threshing by steam were others). The hay-tier or -trusser trade was one route by which independent men elevated themselves above the ruck of dextrous but unspecialised farmworkers.

Hay trussers played a prominent role in Thomas Hardy's *The Mayor of Casterbridge* (1881)—the main character was one of them who had risen to trade in hay and corn. References to the occupation can be found most readily in the period 1880–1918. The Hayward had been a standard functionary in medieval parishes but his role there was to superintend and protect the parish's hay crop rather than bring to market what most villagers anyhow needed for themselves.[33] A document illustrated in Gillian Darley's *Excellent Essex* demonstrates the specialisation which had developed in the environs of London, where demand was always high and market gardening had long been intensive.[34] The image is of an 1860s billhead from the livestock dealers, Frederick Whitbread & Co., Farmers, Graziers and Compressors of Hay, advertised as always on hand to supply agricultural services. The firm already possessed a hay press, a device that had been invented in the United States as recently as 1850 and was a leap forward in making hay easier to store and transport. A further advance came about 1880 when the Dutch barn was introduced, protecting the stored crop from the weather and saving the need to build and thatch hayricks, although back in the eighteenth century Kalm had seen 'hay lathes' with moveable roofs in Ashridge Park.

A description of the hay trussers' trade between about 1890 and 1910 sets out how they proceeded.[35] Hay often stood around in ricks for years until the farmer decided it was a good moment to sell. Hay tiers or trussers then arrived to deal with it. They cut hay from the ricks into chunks, tied them with twine and weighed out half-hundredweights using a manual press. They lived away from their home village for several nights at a time. They worked in pairs, were most active in winter, made a reasonable living, and the remainder of the time did odd jobs. Another account from the same period describes the hay-trussing tasks of a Suffolk farmer's son.[36] His father tested the stacks on offer—farmers liked to sell the whole stack—using a hay-iron by which a sample of the hay could be pulled from the interior. It was then assessed by smell.

The management of hay was already being updated by the mid-nineteenth century, which was when the larger farmers were inclined to specialise or over-specialise in the most profitable grain crops. Richard

Jefferies made it plain that high prices for cereals during the 'Golden Age' led farmers to neglect the ancillary activities that mixed farms had always incorporated, for instance they retreated from the production of pigs, eggs, butter and cheese.[37] Grass and hay were still treasured but a combination of innovative practices had narrowed their function within farm businesses, raised their productivity and replaced part of their role with alternative sources of fodder.

The long-term history of fodder provision is one of steady improvement in the management of grassland, accompanied by the expansion of a range of crops intended to supplement livestock feeding. Among these the most conspicuous were clover and turnips. The role of clover may be slightly over-dramatised in some quarters but, coupled with other introductions, its acceptance into agricultural systems fundamentally reconfigured agriculture. On any one farm and perhaps in certain districts rotations could be altered fairly speedily but the 'fodder revolution' as a whole took the seventeenth and eighteenth centuries to culminate if such mutable arrays as rotations can ever be thought of as finalised. The concept of a 'disruptive technology' is usually applied to industry and regarded as a fairly quick and deep disturbance. During the periods with which we are concerned, farm output and entire agricultural distributions were similarly transformed by a disruptive technology none the less profound for happening in slow-motion.

Notes

1. John N. Hare, 'Change and Continuity in Wiltshire Agriculture: The Later Middle Ages', in Walter Minchinton (ed.), *Agricultural Improvement: Medieval and Modern* (Exeter: Exeter Papers in Economic History, 14), pp. 3–17.
2. Hungerford Local History Group, *Elizabethan Hungerford* (Hungerford: privately printed, 1995), p. 11; F. E. Warneford (ed.), *Star Chamber Suits of John and Thomas Warneford* (Trowbridge: Wiltshire Record Society XLVIII (1992), 1993).
3. Compare Tristan Gooley, *Wild Signs and Star Paths: The Keys to Our Lost Sixth Sense* (London: Sceptre, 2018).
4. John Letts, *Smoke Blackened Thatch: A Unique Source of Late Medieval Plant Remains from Southern England* (Reading and London: University of Reading and English Heritage, 1999).
5. The (purely relative) neglect of the livestock sector is not new. Samuel Rudder circularised the gentry for his eighteenth-century survey of

Gloucestershire but his questions about livestock were simply ignored. David Gowing, 'The Population Geography of Samuel Rudder's Gloucestershire', *Transactions of the Bristol and Gloucestershire Archaeology Society* 141 (1983), p. 148 n. 7.

6. Jeffrey Radley, 'Holly as a Winter Feed', *Agricultural History Review* 9/2 (1961), pp. 89–92; Martin Spray, 'Holly as a Fodder in England', *Agricultural History Review* 29/2 (1981), pp. 97–110.
7. Valerie Belsey, *Discovering Green Lanes* (Totnes: Green Books, 2001), p. 43.
8. John Steane, *Oxfordshire* (London: Pimlico, 1996), p. 79.
9. R. M. Fuller, 'The Changing Extent and Conservation Interest of Lowland Grasslands in England and Wales: A Review of Grassland Surveys 1930-84', *Biological Conservation* 40 (1987), pp. 281–300. I am grateful to Sophie Cunnington of Wild Oxfordshire for drawing my attention to this reference.
10. For examples of fertility covenants see Robin Stanes, *The Husbandry of Devon and Cornwall* (Exeter: privately published, 2008), passim.
11. Violet Howse, *Goosey: A Parish Record Part 2* (Privately printed, 1968), p. xli.
12. A. F. W, Humphrys, *A Few Notes About Marston Meysey* (Privately printed, 1996), p. 6.
13. Vicars Bell, *To Meet Mr Ellis: Little Gaddesden in the Eighteenth Century* (London: 1956), p. 117.
14. The reciprocal traffic in marl and clay between heavy and light land was intended to improve soil texture rather than fertility.
15. Andro Linklater, *Owning the Earth: The Transforming History of Land Ownership* (London: Bloomsbury, 2015), p. 32.
16. Carolina Lane, 'The Development of Pastures and Meadows During the Sixteenth and Seventeenth Centuries', *Agricultural History Review* 28/1 (1980), pp. 18–30.
17. This may explain why the Goose House on Hyde Farm next to Baulking Green, Oxfordshire (formerly Berkshire) has a granary above.
18. Thorkild Kjaergaard, 'A Plant That Changed the World: The Rise and Fall of Clover 1000-2000', *Landscape Research* 28/1 (2003), pp. 41–49.
19. V. G. Simkhovitch, 'Hay and History', *Political Science Quarterly* XXVIII (1913), pp. 385–403.
20. Quoted in E. L. Jones, 'English and European Agricultural Development 1650-1750', in R. M. Hartwell (ed.), *The Industrial Revolution* (Oxford: Basil Blackwell, 1970), p. 60.
21. Jones, 'English and European Agricultural Development', p. 60.
22. G. Eland, *Shardeloes Papers of the Seventeenth and Eighteenth Centuries* (London: Oxford University Press, 1947), p. 61.

23. T. Mozley, *Reminiscences, Chiefly of Towns, Villages and Schools* (London: Longmans Green, 1885), p. 302.
24. E. L. Jones, 'Creative Disruptions in American Agriculture, 1620-1820', *Agricultural History* XLVIII/4 (1974), p. 525.
25. Mark Robinson, 'Archaeological Evidence for Hay Meadows in the Thames Area', Lecture to the Second Hay Meadow Workshop, Wytham, Oxford, 23 June 2007.
26. Cf. J. A. Yelling, *Common Field and Enclosure in England 1450-1850* (London: Macmillan, 1977).
27. S. B. L. Druce, 'An Account of an Embankment and Cutting…', *Journal of the Royal Agricultural Society of England* 2 ser. VI (1870), pp. 367–374.
28. Angus and Owen Davies, 'English Agriculturists' Attitudes Towards Grassland Vegetation, 1780-1914: An Ecological Perspective', *Landscape History* 18 (1996), pp. 71–80.
29. Bell, *To Meet Mr Ellis*, p. 134.
30. J. Mathews, *Remarks on the Cause and Progress of the Scarcity and Dearness of Cattle* (London, 1797), p. 21.
31. E. L. Jones, *Seasons and Prices: The Role of the Weather in English Agricultural History* (London: Allen & Unwin, 1964), p. 103.
32. Frank Emery, *The Landscape of Oxfordshire* (London: Hodder and Stoughton, 1974), p. 109.
33. In the 1990s the office of Hayward survived at Basingstoke, Hampshire. Diana R. Mackarill, 'Haywards in Wessex: The Survival of a Medieval Occupation', *The Hatcher Review* 5/43 (1997), pp. 18–32.
34. Gillian Darley, *Excellent Essex* (Exeter: Old Street Publishing, 2019), p. 11.
35. R. E. Moreau, *The Departed Village* (Oxford: OUP/Readers Union, 1971), pp. 78–79.
36. George Ewart Evans, *The Crooked Scythe* (London: Faber and Faber, 1994), pp. 124–130.
37. Eric Jones, Introduction, *The Farmer's World: Richard Jefferies' Agricultural Journalism in the late 1870s* (Foulsham, Norfolk: Petton Books, 2016), p. xxi. A move back to minor products was evident in the late nineteenth-century arable depression.

CHAPTER 12

Resources: Wool and Wood

Resources and Locations

The doyen of European agricultural history, Bernard Slicher van Bath, once took me to a farmhouse museum in the Netherlands. As we went in, I fell down into a room and doubtless let out an imprecation. 'Ah', said Slicher, 'the loom room, always a metre lower'. Even in the Netherlands, so liable to floods, textile manufacturers felt that extra dampness was worthwhile to stop the threads from breaking. When textiles were made in English cottages, the work was where possible done in a cellar for the same reason. Butter-makers too understood the practical advantages of dampness and sank their dairies three feet into the ground, with the cheese room above. This attention to physical setting is of a piece with the view that the character and distribution of natural resources dictates the location of production. In a typical rendering of this resource determinism, Gwyn Williams tells us that in South Wales, 'the occurrence of coal, iron ore and limestone near each other created the industrial areas'.[1]

The classic case is the Lancashire cotton industry. No one can deny that it rains in that county, whose dampness is often taken to explain its concentrated production of cotton cloth. When geography teachers summarised the matter for their students, they used to take this line—we can call it the 'rising damp' school. Non-specialists found it easy to accept and it did fit (some of) the facts. But by the 1950s, environmental

E. L. Jones, *Landscape History and Rural Society in Southern England*, Palgrave Studies in Economic History,
https://doi.org/10.1007/978-3-030-68616-1_12

considerations began to be derided. The fashion turned to emphasising social and economic factors. I imbibed this version, along with the almost exclusive concern with human decision-making that I came across in history and economics. Academic geographers had long known that things were not so simple and that a conjunction of factors influenced location. Detailed studies rejected humidity as the sole determinant but such complications are hard for teachers to communicate.[2]

How to approach the question of location? Consider the little concentrations of industry in medieval Berkshire, Wiltshire and Oxfordshire. Cloth making had prevailed but gradually shrank away, leaving most of the villages purely agricultural by early modern times. The small towns sometimes developed alternative manufactures and became specialised in one or the other, at least until eighteenth-century transport improvements exposed them to outside competition. Hence Wantage came to concentrate on tanning, Cricklade on gloving, Bampton on leather goods and Aldbourne on bell founding (including a large trade in sheep- and cow-bells). Resource-based explanations are occasionally offered, for instance Wantage is said to have had access to bark from the oak trees of the Vale of White Horse—but other places had similar access. More often local historians leave the issue aside.

When the topic is examined closely, two non-resource factors seem clearer: first, when clusters of enterprises and associated tradesmen emerged they created network economies and could undercut more scattered producers. This helps to explain the concentration of manufactures in a few places within the district, although it does not account for the initial impetus. Secondly, local histories often describe dependence on a small handful of businessmen. This helped particular settlements to rise but its narrow basis made them vulnerable when faced with competition from elsewhere. When the two leading Wantage tanners went bankrupt in the first years of the nineteenth century, no one replaced them, perhaps out of fear they would be unable to compete against the economies of scale secured by their London and Northampton rivals. Similarly, Aldbourne's bell founding depended on one or at most two manufacturers who also went out of business in the early nineteenth century and were bought up by a London firm eager to monopolise the trade.

Bampton leatherwork and Cricklade gloving survived longer, the latter until the 1990s, and it is true that they too depended on handfuls of producers. The market had however become open to all via national transport links and they were unable to expand against the competition of firms in larger towns. Substitutions were hypothetically possible and new enterprises might replace those that had failed, employing the tiny pools of 'industrial' workers in villages or towns that had a history of manufacturing. Yet successor trades like the making of furniture and straw hats in Aldbourne did not last long. Their entrepreneurship was narrowly based. Insofar as any of these occupations used local resources there is no sign that it gave them much advantage; most of the inputs used were widely available.

Discussion of these matters ought to include trade. Suppose local resources ran out or were insufficient to sustain business. Whatever the origins of the industry, it might be continued using resources imported from other districts or even from abroad. When it paid to do so, natural resources could be transported in great loads for long distances, sometimes well before the days of canals, turnpikes and railways. In a fairly extreme example, the Staffordshire Potteries imported clay from as far away as Dorset but the business drive and originality of Josiah Wedgewood were much to the point.[3] Even more strikingly, between 1215 and 1235 15,000 tons of marble had been sent from the Isle of Purbeck to Salisbury, where they were used to raise 3000 columns in the cathedral.[4] This was a luxury item to which ordinary considerations of profit and loss did not apply, but serves to show what the physical possibilities were; the prehistoric transport of sarsens and blue stone to Stonehenge underlines the point. In historical times the cost of transporting stone was greater than the cost of the stone itself or of working it, yet by the early Middle Ages stone was being traded internationally, a trade which became regular from the late Middle Ages.[5] A failure to bring in raw materials reflected not a lack of physical resources but restricted markets that would have made the costs unbearable for commercial enterprises.

Industrial specialisation marked English towns right from the end of the Middle Ages; furthermore the specialisms underwent kaleidoscopic changes. Resource endowments seldom determined the long-run outcome, factors like enterprise, trade and market size being more pertinent. Resource determinism goes out of the window. But there is a catch: the physical world cannot be entirely wished away or I would never have stumbled into the loom room in the Dutch museum. Manufacturing does seem often to have begun where a particular resource was to hand, but if that ran out the options were ceasing business, continuing to grow by substituting resource inputs or switching specialism in mid-stream. In some circumstances, far from universal, hard to specify and dependent on the enterprise of a handful of people, industries became completely detached from their resource base and relied on imports or substitutes, and prospered just the same.

Alternatives to Manufacturing

Questions of resources and location can be looked at via the economic history of the Gloucestershire Cotswolds. The saga is complicated because it involved transitions between industries in different directions. It is not a single case study but a centuries' long sequence of which the main features were the contraction of manufacturing into a subregion in one period and much later the rise of successor industries using quite different resources. In summary, cloth manufacturing concentrated increasingly over several centuries in the Stroud valley, obliging other parts of the county to seek alternatives or go out of business. Their main resort was to move out of manufacturing into commodity trading. Much later, industry in the Stroud valley heartland itself collapsed, leading to a search for other possibilities. Occupations which still employed a physical resource were no longer based on wool but turned unexpectedly to local supplies of wood. Even more unexpectedly this second wave of change touched on the early history of steam-powered industry.

The eastern parts of Gloucestershire had been the first to withdraw from woollen manufacturing. Small towns in the county are seldom so plainly the fossils of medieval industry as are Sherston and Steeple Ashton in Wiltshire but Northleach is an exception. The trade of distributing wool lasted there longer than cloth manufacturing but after the sixteenth century the town fell into deep decay. 'When the trade moved to Stroud and other places on the Cotswold Scarp', Pevsner's Cotswold tells us with perhaps unconscious irony, 'Northleach was left somewhat high and dry'.[6] It was temporarily rescued by the coaching trade. The burgage plots of Northleach are traceable on the ground and a court leet continues to be held. Otherwise most of East Gloucestershire became agricultural, with only sporadic efforts at finding other activities. Most conspicuously, some of its water mills switched to paper-making, yet that was a trade which did not absorb many hands. The area never returned to cloth making and stayed almost entirely rural.

With respect to the first wave of alterations, consider Cirencester. In the Middle Ages cloth making had been its common occupation, as in so many southern towns, but when this was outcompeted it created an opening to specialise in trading in wool: Cirencester's leading inhabitants became wool-staplers, a term sometimes used interchangeably with its best paid subspecialism, wool sorters, who did the hard and often dirty work of cleaning, grading and packing the raw material. The staplers proper were those who invested and dealt in stocks and supplies of wool. Cirencester thus made a highly satisfactory transition to becoming an entrepot to which wool was brought from astonishing distances for a raw material—as far away as Lincolnshire and Pembrokeshire—processed by the wool-staplers and sold to the cloth manufacturers of the Stroud valley. Local Cotswold wool was less relevant because its quality was comparatively poor and the volume available did not equal the enormous loads attracted from distant parts of the country.

At the start of the seventeenth century, Cirencester was already the least specialised town in Gloucestershire and was a place of traders.[7] The competition that had overturned the town's woollen industry came from Stroud valley mills benefiting from stronger water power. The historian of medieval Cirencester, David Rollison, thinks the town's industry shrank because of the lack of rivulets on which mills might be sited. The limited power generated at the existing mills was insufficient for them to raise their output (and cut their prices) to compete with the better-endowed Stroud valley mills. It is at least as likely that the profits from becoming a mart for wool actually leached capital and enterprise from the town's comparatively marginal manufacturing enterprises, meaning that Cirencester may have been as much drawn out of industry as pushed out. Business dealings are much less visible and less easy to record than physical features of mill sites and streamflow, which means that while local historians may mention commerce they seldom delve into it deeply.

In Cirencester, wool-stapling may have begun to eclipse cloth making as early as the mid-fourteenth century. The town came to acquire and sell wool on a vast scale, not only to Stroud but also to North Country wool-brokers for the Yorkshire worsted manufacturers. It developed expertise in the 'soft' skills of trading and marketing. New businesses employed numbers of carders, combers and so forth to prepare the raw material. Who was responsible for this prolonged but powerful transition is not clear—whether manufacturers who had some capital and had seen the writing on the wall or completely fresh business entrants. Fortunately for Cirencester it was not hard for newcomers to gain a footing, given that in the sixteenth century the town had no guild strong enough to confine trades to existing residents. Maybe a big part of the shift was brought about by Quaker and nonconformist enterprise, since those sects included prominent traders with nationwide contacts.

Wool-stapling was obviously a paying proposition and even in the late nineteenth century wool stores were still being built in towns and villages across Southern England. But for a whole town to grow successfully by replacing its manufacturing by trading on a large scale was distinctly unusual. Cirencester and neighbouring Tetbury were the most successful in the region. No wool marketing building survives in Cirencester but Tetbury market house was built in 1655 as the 'Great Tolsey' for the sale of wool and yarn. These were not the only towns trying to capitalise on trade in wool, just the most effective. The wool store built at Bibury in the fifteenth century was converted into the cottages of Arlington Row as early as the seventeenth century. Minchinhampton's historian is sure that his town originally had three market houses, all intended for storing wool.[8] The third and architecturally most impressive one is still extant. It was built by Philip Sheppard in 1698 as a mart for wool and yarn in a great clothing country—that is, to supply woollen manufacturers. The building survives but was never a commercial success.

Neither Cirencester nor Tetbury had the 'site advantage' of natural resources: their water power was weak compared with the Stroud valley. Their labour was not cheap enough to offset the power from fast streams or later to compete with coal-fired steam in the north of England.[9] Yet Cirencester and Tetbury did have a 'locational advantage'. Both were close to sources of demand in the Wiltshire and Gloucestershire cloth industries and inherited reasonable road systems along which distant, especially Midland, wool could come. The main impetus for change was probably entrepreneurship but this is difficult to fathom from the available evidence. It is easier to understand in the case of Chippenham, Wiltshire, where the answer does seem to have been individual enterprise inspired by political ambition.[10] The Fludyer family arrived there and became great factors in the wool trade. Why so few other clothing towns reinvented themselves as specialist entrepots is not clear. Poor access to manufacturing areas with a high demand for wool is one possibility. The cloth trade in Newbury, for instance, declined during the seventeenth century with no sign that it might turn to trading in wool. The town has a cloth hall (now a museum), a corn hall and a corn store but not a wool store. It was simply not close to the surviving woollen manufacturing areas.

Specific details of Cirencester's trade are hard to come by, although the records of Ditchley Farm, Oxfordshire, note that on 4 November 1786,

£122.9 s.0d was received from William Newcombe of Cirencester for 149 tods of wool (4172 lb).[11] Men in distant Yorkshire were however aware of Cirencester as a significant centre of wool-staplers. I own an account book of 1829–1830 from the marvellously named Idle in the West Riding which carefully notes the address of a Cirencester wool-stapler whose dwelling house, premises, counting house and mill were at the north end of Gloucester Street. This is now The Old House which is still standing in Gloucester Street and he was Thomas Slatter, who was also a tanner and a fellmonger dealing in sheepskins, as well as being active in local politics.

The wool brought in large volumes from the Midlands and much further afield was meant to supply the manufacturers of Wiltshire and Gloucestershire, most of whom collapsed quite fast about 1830, undermining Cirencester's function of warehousing and redistributing the raw material. In any case the town's wool trade was already shrinking in the late eighteenth century. Buyers had begun to short-circuit the market by riding out to farms to strike individual bargains. Gloucestershire clothiers even began to travel to Saxony for higher quality wool and were quick to tap Australian supplies. These developments amounted to disintermediation but are not presented as a linked sequence in local histories. They influenced the fact that Cirencester's population growth was already slow in the eighteenth century. During that period of national expansion such sluggishness was a come-down for what had been an active and prosperous place: 4000 people in 1700 but only 4130 a century later.[12] Cirencester's actual manufacturing had ebbed gradually but by contrast the second transition, the abandonment of textile marketing trades, was quite abrupt: 13 per cent of the 600 householders entitled to vote were employed in textiles in 1790 but only nine per cent by 1812. There were 34 wool combers in 1790 but only 17 by 1812. By 1822 Cirencester had more attorneys than wool-staplers. The population reached 8431 by 1881 but this does not compare with growth in the northern industrial cities.

On the basis of the Cirencester example it might be decided that a local economy had no need to found its manufacturing on the availability of raw materials or power. Trade could be an effective substitute. Insisting on the necessity of a physical resource base would be incorrect. Nevertheless it would be unwise to argue the opposite—that trade or substitution could or would always replace absent or exhausted natural resources.

Successor Industries

Studies of the economic history of Gloucestershire seem disconnected rather than generalised. The business history of adjustments to unfavourable circumstances is rarely treated *en bloc*. The first substitute for cloth making, wool-stapling, ceased in turn, but discussion of what replaced it is restricted, at least in Cirencester's case, to descriptions of its 'decline' into an agricultural market town and later emergence as a leisure resort catering especially to fox hunters. It even became known as the Melton [Mowbray] of the West. And in Gloucestershire's collapsed manufacturing heartland of the Stroud valley the alternatives—the 'successor industries'—are dealt with fragmentarily.[13] Most historians understandably discuss the demise of the large mills at Stroud which, although some had adopted steam power, succumbed quickly to competition from Yorkshire. A diverse range of industries was tried in efforts to repurpose the mills: the emphasis was on reusing premises that were sturdy and had been expensive to build. Yet focussing on the closure and reuse of mills in the Stroud valley itself in Victorian times truncates the story. Areas to the north and south had long been obliged to adapt, as had wider areas of East Gloucestershire, to Stroud's pre-eminence in cloth making. They had done so, first, by trying to move from industry to trade after the fashion of Cirencester and Tetbury and later by turning to products largely alien even to the successor industries of Stroud.

The tendency in the literature is to tackle individual aspects of the woollen industry rather than the downturn as a whole, for instance to discuss labour supply, poverty or emigration. When a large fraction of the workforce was jobless for any length of time its experiences were like those of the self-employed who suffered from the 'handloom weaver problem', indeed their situation was worse because independent weavers were able to continue making a few pence in their cottages for a while whereas unemployed mill hands had no such outlet. Their plight reflected an inability to find other employment or quit the district—or the failure of society to help them adapt. Efforts were made to promote overseas emigration but the details of who left, when they did so and precisely why, are often enigmas.[14] Wage labour was almost collateral damage from the decline of an industry whose costs proved too high in the face of falling product prices or whose employers were guilty of timorousness or mismanagement. These issues are mentioned in the literature but what happened to the failed employers themselves is thoroughly obscure.

Viewed on a timescale of centuries, changes in industrial distributions have been obscure yet nevertheless mutable. To emphasise the point it is worth mentioning some of the phases and distributions. During medieval times some cloth making, once heavily centred in towns, had expanded into the villages. Meanwhile Gloucester's cloth industry declined and Stroud's rose, while during the eighteenth century the Wiltshire industry retreated in favour of Gloucestershire. Two separate authors imply that workers 'invaded' the Gloucestershire weaving trade when they were put out of work during more than one agricultural depression or when land-use changes ousted farm labour or made it redundant.[15]

Most villages and small towns in the agricultural districts retained only small tradesmen and craftsmen who could make a living by supplying their neighbours or repairing equipment. These enterprises did not grow much and it has to be assumed that the failure of such activities as carpentry, malting and basket-weaving to expand was due to the limited size of the market, meaning small populations and poor communications. For a time the cost of transport protected local craftsmen in these trades, although during the nineteenth-century firms in distant towns, even overseas, began to undercut them.

Sometimes there was an intermediate phase when craftsmen turned to buying and incorporating factory-made inputs until at length the entire product could be brought in more cheaply. This happened to the wheelwrights and clockmakers. The latter were 'hollowed out' to the extent that they began to buy in the mechanisms but paint their own names and towns on the dials as if they still made the entire clock. When cheap clocks from Connecticut became available in the nineteenth century the market in everyday timepieces was lost. Local clockmakers survived mainly by selling pricey long-case (grandfather) clocks or by repair work. A few managed to shift into crafts that bore no direct resemblance to their original trade. They capitalised on sectors of growing demand which utilised their basic technical skills: thus the Honeybones in Fairford shifted from the clock trade to fabricating fishing tackle for the well-heeled visitors who came to fish the river Coln. An unexpected parallel is how in Japan, after the collapse of Tokugawa demand for armour and hand weapons, the producers of these switched to making tackle to supply the craze for fishing under Meiji.[16] Similarly, the prominent bell founding activity at Aldbourne, Wiltshire, seems to have been evolved by a gunsmith in the Corr family, who finding his trade reduced after the Civil War and Commonwealth periods, sought an alternative metal industry, i.e.

bell founding. An archaeological dig during the twenty-first century has unearthed a small bell made of gunmetal, which may signal the transition from one activity to the other.[17]

Some people in affected or distressed districts thus moved sideways into other businesses. Almost idiosyncratic options were sometimes chosen—for instance Anstie, the leading woollen manufacturer in Devizes, Wiltshire, moved into groceries and snuff. He was aided by his connections with the Bristol tobacco trade, demonstrating the role of individuals and indeed of happenstance.[18] Alternatively, the possibility that a local attraction would draw in customers might do the trick. A geographically distant example was the evolution in the use of wherries on the Norfolk Broads. They had been an important mode of freight transport until they were undercut by the mid-nineteenth century extension of the railway network. In a creative response, wherries were converted into pleasure boats for hire by holiday-makers, first seasonally, then permanently, with the traditional black sails changed to white ones. Subsequently wherry-yachts began to be built. There seems as yet no common explanation of why some shifts succeeded and others did not. In small places a major factor was probably the presence or absence of one or two enterprising families.

Returning to our main theme of cloth making, communities which during the late medieval and early modern periods could not compete with the greater water-power of the Stroud Valley had a number of options, although not all were available at the same period. Transitions were slow when transport was expensive, markets were not fully connected and prospects uncertain. Towns outside the valley tried to adjust to the difficulties of their weavers, trading in raw wool being the commonest move. It is hard to persuade many people that any mention of wool does not necessarily imply weaving it into cloth. Theirs is the Materialismus mindset which thinks that making physical commodities (in *The Economist's* phrase, things you can drop on your foot) is ipso facto superior to service sector transactions or any business that shuffles papers. 'Superior' conjures up something more mystical than mere profitability. Trade and office work were less visible than manufacturing and did not bequeath the tangible relics of tall mills and sturdy water wheels; they were like performance art and when commercial businesses ceased, important though they could be, it is as if they had evaporated.

The late fourteenth- and early fifteenth-centuries had been the times when wool-staplers rose to wealth and prominence and built the

Cotswold 'wool churches'. The churches rose on the basis of receipts from trading, not from manufacturing. In terms of profits Cirencester did best of all. Whereas the town lacks woollen mills or a pillared market hall like its rivals, several buildings representative of its trading function do survive, including a tallymans's premises, a warehouse and the counting house of the warehouse's owner. The town's prowess was widely known. But when manufacturing declined in the Stroud Valley there was no longer a need to amass large volumes of English wool. As noted, Saxony (which provided 65 per cent of imported wool by 1820), followed by Australia, was already supplying the market, undercutting the Cirencester trade.[19]

Trading in wool had been profitable for a long time but its eventual shrinkage showed that it was not after all destined to be a permanent alternative to cloth making, as must have seemed the case for centuries. What, then, were people to do? Bankrupt wool-staplers perhaps left the district. What really happened to them is unknown, just as it is with respect to farmers in arable depressions—it is rare to be able to trace those who left agriculture. They almost dematerialise. A clue might lie in the testimony by J. B. Turner of Brockhampton Hall, Herefordshire, to the Select Committee on Agriculture in 1833.[20] He stated that farmers who quit during the post-Napoleonic agricultural depression took whatever job they could find, becoming farm bailiffs, schoolmasters or clerks. Some may have fallen on parish relief or the charity of friends and we can guess that some emigrated, taking them out of the reach of local research. Their exodus is probably a model for people moving out of making cloth or trading in wool.

The Wood-Working Alternative

Once cloth making had virtually collapsed and carried down with it the related activity of dealing in wool, nineteenth-century Gloucestershire was left with two main options. Industry and related trades could be relinquished, leaking both labour and employers, and seeing the districts concerned become purely rural and losing much semblance of widespread prosperity. The other was to move into different manufacturing pursuits for which there might still be a market with less competition; of these the most interesting was wood-working, which returns us to the topic of resources in the sense of raw materials.

The western fringe of the Cotswolds was thick with beech trees, somewhat pretentiously called 'the weed of the Oolite'. The woods stretched from Birdlip to Wotton-under-Edge. The southerly distribution of the species in England, coupled with the opportunity for transporting it along the Severn Valley, made the western Cotswolds a convenient source for the Birmingham gun makers. No other large supplies of beech were within their reach and they needed it for cheap gun-stocks destined for the huge African slave trade. This was especially the case once the countryside had been denuded of the preferred alternative of walnut during the French Revolutionary and Napoleonic Wars. Walnut was much preferred for gun-stocks because of its attractive grain but had become unobtainable in anything like the volume required. It had then to be imported from overseas, despite efforts to plant walnut trees explicitly for the purpose. There was even a committee formed in 1804 for that purpose at Evesham, Worcestershire, during the panic about Napoleonic invasion.[21] For the stocks of cheap guns, rough-cut blanks of beech were acceptable, although it was considered an inferior wood because the lack of pattern in the grain reduced its strength. This switch between resources will not surprise economists, who expect substitutions to take place and put little credence in laments about adamantine physical scarcity.

At the end of the eighteenth century dealers were already converting some Gloucestershire woodland timber, presumably beech, into blanks for gun-stocks. Once the blanks had been 'deprived of waste bulk' they were sent to Birmingham. The residue and lesser pieces of wood were made into charcoal. Birmingham is known to have obtained timber from Herefordshire and Gloucestershire, unfortunately with few details forthcoming.[22] Intriguingly, advertisements for timber sales in Herefordshire and Gloucestershire are annotated in copies of the *Gloucester Journal* for 11/4/1825 and 26/3/1831 which are held in the archival collection of the Bagshawes of Norton, then in Derbyshire, now in Sheffield.[23] This family, which had interests in woodland elsewhere, owned timbered property at Parson's Hill, Sudgrove, in Miserden parish on the Cotswolds. It is tempting to speculate that the family formed the link connecting sales of beech staves to Sheffield and Birmingham to make packing cases in which manufactures were exported. On the Sandys estate at Miserden the chief watermill was used to make staves for this purpose, representing a reversion in technology from the local use of steam. This mill lasted only through the 1830s. It had already been converted from corn milling to flock-making, though that had not lasted either.

Away from Stroud, in smaller places such as Nympsfield, Avening, Minchinhampton and above all Nailsworth, workshops sprang up to make gun-stocks, furniture, bed frames and walking sticks, and were in operation for at least the first forty years of the nineteenth century.[24] This was handicraft industry in workshops which, even if they are still standing, are unlikely to display specific traces of their use. The fabrication of blanks for gun-stocks which took place on the woodland floor will have left none. As early as the 1770s, gun-stocks had been made from beech in woods to the north of Stroud, specifically at Miserden and Cranham, as Samuel Rudder had noted.[25] By 1820 a steam engine was employed at both these places. There were four steam engines in the Gloucestershire woollen industry in 1815 but this period was early for steam in a wood-working industry; even in London steam took over from hand-sawing in pits only from the 1840s.[26]

The clue to the early use of steam power in the rather remote woods of the Cotswolds was chancing on the phrase 'steam engine' marked on Bryant's map of Gloucestershire (published in November 1824) along a tributary of the Frome on the Miserden estate. The words are not shown on other maps of the same period although three buildings occur on later nineteenth-century ones and less often the phrase 'Engine Bottom' or 'Engine House' appears, just below a 'Fish Pond'. The site is shown best on an 1874 plan of the estate but is not marked or named on modern maps.[27] This find, forgotten even by residents of the estate, led me in a roundabout way to discover the existence of a wood-working industry there, besides south of Stroud, as one of the successor industries to cloth manufacturing.

The Miserden estate was owned by Sir Edwin Baynton Sandys who became entangled in debts and financial dealings which obliged him to sell the estate in the early 1830s. His problems are detailed in the Sandys collection in a document relating to a case brought by his creditors.[28] The document is utterly convoluted and unorganised. The legal recounting of the arrangements into which Sandys had entered covers 213 handwritten pages. Just enough passing mentions occur to reconstruct the following narrative: Sandys had become involved with the timber merchants, William Todd and his partner, Frederick W. Courthard. The latter was from Canada Wharf, Rotherhithe; presumably both were. Todd and Sandys were friendly enough at the outset for Todd to stay in Sandys' house. In 1818 Sandys agreed to let to Todd land at Ladlecombe,

Cranham, on which to build himself a villa using timber from the estate, although Todd was also felling timber for commercial purposes.

Todd and Sandys were plainly birds of a feather. The latter came down hard on the labouring classes in defence of his property while Todd made himself thoroughly unpopular in Cranham by encroaching on and enclosing common land, and failing to pay the compensation demanded by the Court Leet.[29] He built what amounted to a small holiday village, expensively appointed and much visited by fashionable people from Cheltenham. He put up several cottages and outbuildings, 'together with certain Steam Engines Saw Mills and other erections for the purpose of manufacturing and preparing the Timber'. Early steam engines sometimes used charcoal as fuel and there was plenty of that made at Cranham. Histories of Todd's tourist resort play up art, architectural and literary aspects and ignore the basis of his operation, which was felling timber and using it in manufacture.

Todd and his partner both went bankrupt in 1822 but the former was accused of continuing to fell large quantities of estate timber, i.e. beech and a lesser volume of ash. He was still 'speculating' in local beech timber in 1826 and did not leave Cranham until 1828. The assignees to whom money was owed had earlier felt themselves, 'to be at liberty to move the Steam Engine' and are soon reported to 'have also removed said Steam Engine from said premises'. The suspicion arises that the engine from Cranham became the Miserden one. Conceivably Sandys himself was among those seizing it because he was not obliged to sell his estate until the 1830s, although no reference is ever made to a steam engine as an estate asset.

Shifting an engine into an out of the way valley bottom in Miserden Park would not have been easy but installation would have been difficult wherever it took place—as a Leeds industrialist said of early examples, 'the steam engine was about as portable as a parish church'.[30] Portable steam engines were not yet available. The first was made in 1839 at Tuxford, Nottinghamshire, interestingly at the suggestion of John Morton (Gloucestershire agent for the Earl of Ducie) who recommended putting small engines on wheels.[31] The chronology of steam at Miserden Park is unclear but the survey for Bryant's map which locates an engine there had been carried out in 1823 and 1824. Whatever the details, it is evident that steam was introduced at Cranham and Miserden at early dates in order to work up timber for sale. The aim was to mechanise the manufacture of gun-stocks of beech for the Birmingham market. Trade directories for Birmingham about 1830 list many gun-stock makers, besides a specialist walnut timber dealer and even a gun-stock varnisher.[32] Fifty-one different gun-making trades were present in Birmingham in the early nineteenth century. A division of labour among gun manufacturers had been evident by 1767 and is not surprising given the vast scale of manufacture. By 1790 over 4000 gunmakers were despatching 100,000 guns per annum to slave traders. These guns were dangerous, not being 'proved', and the precise methods of their manufacture seem impossible to trace.

Gun-making is usually said to have resisted mechanisation until the 1850s, when machinery to cut out stocks was at last imported from the United States.[33] The assertion accepted by historians of technology, that English gun-making methods had been adapted to the requirements of individual customers and inimical to mass production, is misleading.[34] When the market became big enough, as it did with the huge demand for the African slave trade, it encouraged partial mechanisation in the unfamiliar setting of the Gloucestershire woods. The wood-working industries there have not been recognised because the published history is sundered into that of the individual parishes and because of the misleading assumption that industries need big buildings. Nor is much distinction made in the literature between small craftsmen supplying the local market and those who produced for sale outside the district. Insofar as Cotswold craftsmen made furniture, they were eventually undercut by London makers and beat a retreat into producing fashionable bespoke items.[35]

Other Options

The decline of Stroud's woollen industry was protracted. Hopeful new entrants appeared even in the 1830s and 1840s. A tiny proportion of mills experimented with new fabrics and managed to make cloth throughout much of the nineteenth century. Nevertheless by the second quarter of the nineteenth-century Stroud was no longer a major industrial centre with a defined role. Redundant mills went through a sequence of other uses. Some of them were still linked to wool, notably dyeing and the production of shoddy (an inferior fabric made from waste wool). Then, as supplies of waste from the local cloth industry dried up with the decline of cloth manufacturing, waste had to be brought in from elsewhere—once again a resource scarcity was overcome by trade. Silk manufacture was also introduced, as were pin-making, the fabricating of fibre- and other types of board, and as a last resort humble corn milling. Stephen Mills emphasises that any redundant cloth making mill did at least offer space, an open plan layout and a source of power. The buildings were solidly built and spacious enough to be shared by more than one firm and there were plenty of mill buildings, 200 having been erected almost in a frenzy between 1800 and 1820.[36] Hence the collapse of woollens was not totally without substitutes, although the many messy transitions and their overall inadequacy as replacements for the traditional industry are hard to summarise in a single phrase.

Trades using wood were prominent among what came in. Several mills became saw mills, made cabinets or went in for wood turning. From at least the 1840s the largest single novelty was making walking sticks, which in the second half of the nineteenth century occupied at least seventeen mills. This does not sound a significant industry to modern ears but contemporary fashion made it more important than it seems. It began by using the abundant and cheap local beech wood but over the years took to importing a wide variety of foreign timber, in another instance of different materials being substituted for the original resource.

The previous paragraphs on 'successor industries' are drawn from Stephen Mills's thesis. He is critical about errors and omissions in previous gazetteers of mills, justifiably up to a point because his interest is in what happened to cloth making in Stroud, an industry which was physically situated in mill buildings. Nevertheless, despite Mills's preference for archaeological techniques over documentary sources, only the two approaches combined will permit us to discover the whole calculus underlying industrial location and relocation. Either type of source on its own leaves gaps in the story. An over-emphasis on the side of 'heritage' and architectural remains leads away from economic and business history.[37]

To grasp how businesses and economic institutions worked means studying their function rather than relying on architectural surveys that might unkindly be termed outdoor museology, especially when the focus of attention concentrates on structures which are still standing. Only a broad approach that includes trade, technology and market demand serves to identify industries which spawned no businesses with large archives. Focussing on mills and mill buildings has distracted attention from dispersed activities that took place in workshops (some perhaps little more than back-garden sheds). It is particularly unable to detect the preparation of gun-stocks in situ on the floor of beech woods where there were no buildings at all. The region's forgotten cluster of wood-working industries should be considered in its own right. Equally, the presence of beech along the Cotswold escarpment deserves to be understood as a founding, though impermanent, influence on the relocation of production.

Notes

1. Gwyn Williams, *The Land Remembers: A View of Wales* (London: Futura, 1978), p. 187.
2. Wilfred Smith, *An Economic Geography of Great Britain* (London: Methuen, second edition 1953, first published 1949), especially p. 166. Smith gives considerable detail about factors influencing the location of the cotton industry.
3. See Eric L. Jones, 'The Fashion Manipulators: Consumer Tastes and British Industries, 1660–1800', in Louis P. Cain and Paul J. Uselding (eds.), *Business Enterprise and Economic Change* (The Kent State University Press, 1973), pp. 198–226.
4. Christopher Hadley, *Hollow Places* (London: William Collins, 2019), p. 164. Under Norman rule the church was powerful enough to ensure that the costs of this stupendous carriage were met.
5. Andrezej Wyrobisz, 'Resources and Construction Materials in Preindustrial Europe', in Antoni Maczak and William N. Parker (eds.), *Natural Resources in European History* (Washington, DC: Resources for the Future, 1978), pp. 74ff.
6. David Verey and Alan Brooks, *The Buildings of England: Gloucestershire 1: The Cotswolds* (New Haven: Yale University Press, 2000), p. 512.
7. David Rollison, *Commune, Country and Commonwealth: The People of Cirencester, 1117–1643* (Woodbridge: Boydell, 2011).
8. Arthur T. Playne, *A History of the Parishes of Minchinhampton and Avening* (Gloucester: John Bellows, 1915), p. 116.
9. The Stroud Valley did at length introduce steam engines but still failed to retain much of its woollen manufacturing. The blanket manufacture at Witney, Oxfordshire, was the regional anomaly in relying on cheap labour to overcome its disadvantages of little water power and no coal.
10. Richard Baines, *A History of Chippenham* (Chippenham Civic Society, 2009), passim.
11. Oxfordshire Record Office, Ditchley Farm Account, [1781–1786] DIL I/c/22a.
12. These and other figures come from various sources, but see especially Rollison on the earlier periods. Later times are described in detail by K. J. Beecham, *History of Cirencester* (Stroud: Alan Sutton, 1978 [1887]).

13. The striking exception to a typically incomplete coverage is the almost overpowering detail in the 1997 Leicester Ph.D. by Stephen Mills, The origins, development, decline and reuse of the Cloth Mills of the Stroud valleys of Gloucestershire: A Study in Industrial Archaeology. It was posted on the Internet in 2014 at Leicester.figshare.com > articles > The _origins_development_decline_... Mills contains a critical review of the literature, pointing out that little had been written on 'successor industries', the reuse of buildings or the experience of the labour force. He concentrates on those topics and on technology. Another paper (R. Perry, 'The Gloucestershire Woollen Industry', *Transactions of the Bristol and Gloucestershire Archaeological Society* 66 (1945), pp. 49–137) presents similarly exhaustive detail for the whole county but does not deal with 'successor industries'.
14. Most information is available about the late 1830s emigration of 68 people from Bisley under the sponsorship of Rev. Thomas Keble. David Ricardo, who lived at Gatcombe, wrote a commendation of this transfer of labour from Gloucestershire to New South Wales.
15. Perry, 'Gloucestershire Woollen Industry', p. 94; Edith Brill, *Cotswold Crafts* (London: Batsford, 1977), p. 171. Neither author is clear about the chronology or scale of this movement. A reverse shift of weavers into farm work was said not to be possible because weavers lacked enough stamina.
16. Eric Jones, 'The History of Biological Exploitation on the Pacific Rim', in Inderjit Kaur and Nirvikar Singh (eds.), *The Oxford Handbook of the Economics of the Pacific Rim* (Oxford: Oxford University Press, 2014), pp. 48–49.
17. Ex inform, John Dymond and Terry Gilligan, Aldbourne Community Heritage Group.
18. Lorna Haycock, *John Anstie of Devizes 1743–1830* (Stroud: Wiltshire Archaeological and Natural History Society and Alan Sutton, 1991).
19. Lionel F. J. Walrond, 'Wool, Woolmen and Weavers', in Charles Hadfield and A. M. Hadfield (eds.), *The Cotswolds: A New Study* (Newton Abbot: David and Charles, 1973), p. 184.
20. Quoted in Eric Jones, *Landed Estates and Rural Inequality in English History* (Basingstoke: Palgrave Macmillan, 2018), p. 25.
21. R. C. Gaut, *A History of Worcestershire Agriculture and Rural Evolution* (Worcester: Littlebury & Co., 1939), p. 298.
22. G. C. Allen, *The Industrial Development of Birmingham and the Black Country 1860–1927* (London: George Allen & Unwin, 1929/Routledge 2018), p. 102. It was stated that 'considerable quantities' of timber were being sent by canal from Herefordshire to Birmingham in 1838 but species and volumes are not indicated. *Report* by Stephen Ballard on the Gloucester-Hereford canal, 1838 (Hereford Paving, Canal and Road Acts,

Hereford City P.C. 626). A 'careful estimate' in 1849 was that 10,000 tons of timber was felled annually in Herefordshire (*Hereford Journal* 18 July 1849).

23. Sheffield Record Office, CD1304. I am grateful to the Archivist, Robin Wiltshire, for useful clues concerning the Bagshawe family.
24. See British History online for the respective parishes; Alan Tucker, *Traffic on the Stroudwater Canal: 1755–1822* (Gloucestershire Canals Restoration, 2007); Cyril Turk, 'A Revolt Against the Rates in Minchinhampton', *Minchinhampton Local History Group Bulletin* 2 (1985), p. 5; and Thomas Rudge, *General View of the Agriculture of Gloucestershire* (London: R. Phillips, 1807), pp. 240, 245. Rudge's wording hints that the sale of local beech for making gun-stocks, together with furniture, was more widespread than the parishes named.
25. Samuel Rudder, *A New History of Gloucestershire* (Cirencester: Privately printed, 1779), pp. 396, 553.
26. Joanna Smith and Ray Rogers, *Behind the Veneer: The South Shoreditch Furniture Trade and Its Buildings* (Swindon: English Heritage, 2006), p. 16; Walrond in Hadfield, *The Cotswolds*, p. 188.
27. Documentary sources used are to be found in the Sandys Family of Miserden collection in the Gloucestershire Archives, D1042/Box 1.
28. Direct quotations and other material relating to steam at Cranham are drawn from the Sandys collection in Gloucestershire Archives, D1042/Box 1/9.
29. Janet Whitton and Kate Searle, *Cranham: The History of a Cotswold Village* (Privately printed, 2005), pp. 164–165, 171, 175.
30. Quoted by Kenneth Brown et al., *British Road Steam Vehicles* (London: Collins & Brown, 1999), p. 5.
31. National Museum of Scotland website, 'Tuxford Portable Steam Engine'. Other manufacturers were slower off the mark although the blog of Market Lavington museum, Wiltshire, hints that William Cambridge made portables there in 1837.
32. William West, *The History, Topography and Directory of Warwickshire* (Birmingham: R. Wightson, 1830).
33. See e.g. the Foreword to the website of The Birmingham Gun Barrel Proof House Museum.
34. Edward Ames and Nathan Rosenberg, 'The Enfield Arsenal in Theory and History', *Economic Journal* 78/312 (1968), pp. 827–842.
35. Brill, *Cotswold Crafts*, pp. 86–87.
36. Walrond in Hadfield, *The Cotswolds*, pp. 192, 196.
37. The analysis in Smith and Rogers, *Behind the Veneer*, is similarly hampered by restricting itself to building history.

PART III

Conclusion

CHAPTER 13

Conclusion

The core principle of this book is that landscapes and natural history offer guides to historical processes that otherwise are easily overlooked. Consider again previous cultivation out on the heaths, as discussed in Chapter 6. Prehistoric archaeologists might be thought to have discovered this but they missed the incursions of the plough in the New Forest which Colin Tubbs and I found through noticing regularities in the masses of gorse. The possibility that such sites of cultivation might exist in numbers was dismissed, as is the way when professionals are told about things they 'know' can't be there, in this case illegal cultivation on Crown land. The sites we found did not make a large addition to the nation's cultivated area—marginal cultivation is by definition marginal—but they did suggest the pressure on the land at sensitive historical periods. The discovery set us looking for confirmation in libraries and archives. At other times the opposite happened and it was library work that first revealed some unknown feature of the landscape.

The investigations, visits and researches described have been enjoyable, even if academically that might be thought the least respectable of motives. Walking the fields and streets, talking to old countrymen, scouring second-hand bookshops and visiting the tattered Local History sections of small-town libraries has been a lucky dip where the discoveries often surprisingly knitted together. The more respectable motive has been

E. L. Jones, *Landscape History and Rural Society in Southern England*, Palgrave Studies in Economic History,
https://doi.org/10.1007/978-3-030-68616-1_13

trying to exploit landscape observations when writing about economic and agricultural history more generally. The findings have fed into my academic studies and, starting as musings about something noticed on the ground, have occasionally been elevated to an almost global scale.[1]

From this it can be seen that what began as spare-time rambles were not originally inspired by any meta-theme or grand theory. They began as hobbyist explorations. One of the pleasures of the English countryside is that few minutes in the field need ever be wasted: natural history can be carried on side-by-side with local or landscape history. My personal friends have tended to be people who share the same joint interests in natural history and local history—a select band several of whom have ended up publishing on similar themes. For guiding ideas, I have borrowed or stolen from my professional contacts, starting on field trips with the exceptional geography master (and geologist) at my school, W. D. Dreghorn, continuing at the research student stage with the economic historian, Maurice Beresford, and culminating in a lifetime investigating the landscape whenever I could escape from my desk. This book is a report on some of the results.

In my undergraduate days it took me some time to realise just how contingent landscape history can be. In my case that meant the history of the Hampshire chalk. I was misled, as conservationists often still are, by the attractiveness of small patches of down-land that had escaped cumulative extensions of ploughing. Before myxomatosis put a stop to grazing by rabbits, grassy downs dotted with some, but not too much, hawthorn scrub were rich in birds, butterflies and orchids. This happy conjunction was easy to accept as the baseline, the Eden from which the countryside fell into what Mark Cocker calls the 'cavernous heart' of England, endorsed as a biodiversity desert by Dave Goulson, another of the leading naturalists writing today. Eager volunteers nowadays spend their time battling the spread of scrub on surviving chalk land fragments, trying to stabilise the habitat I saw as a boy. Those I have talked to speak as if this was the natural state of affairs. In the back of their minds they do understand that the landscape has passed through other stages, without showing much interest in the history or the fact that the open downs were artefacts of grazing by sheep under particular price regimes. The grassland economy is unlikely to return, short of a cataclysmic economic and demographic crash, and only rear-guard action can preserve bits of its ecosystem in any earlier condition. What I saw, and what conservationists now aim at, amounts to a single phase when parts of the chalk land were slowly going back to scrub. Naturalists may wish to arrest change at that stage and have had some partial success, yet restoring a historically transient artefact is labour intensive and thoroughly arbitrary. More thought should be given to what habitats are desirable, and why.

Ground evidence can be mesmerising. Excellent though it is in providing clues to neglected topics, one should keep one's mind on their economic purpose. An example is the drove road. Because it can be a captivating pastime to identify droves and green lanes on the ground, the visible evidence seems to monopolise attention. People look for signs of tracks on early maps and in published sources and are distracted from investigating the function of the routes, which means the volume and value of traffic they carried. These are almost intractable topics—the landscape is silent about usage—but a little headway may be made through reports of the fairs where livestock were sold. Total receipts nevertheless remain hard to calculate. On probing, however, it unexpectedly emerges that, while spectacular movements of Scottish and Welsh beasts were regular right from Skye or Anglesey to the London area, the relative importance of long-distance droving has been exaggerated.

Landscape evidence can thus direct one to themes and thoughts which might otherwise escape notice but is a supplement rather than a substitute for library and archival work. All sources are likely to contain gaps, some of them as wide as a barn door and the record inscribed in the landscape is no exception. Certain activities have left far fewer traces on the ground than others, which may lead to distortions of the type evident in studies of the woollen industry. The weaving of cloth in cottages, though well-known, has left relatively few spoor; it could hardly be otherwise when so many of the humbler cottages of the past have been pulled down. They were seldom designed for home manufacturing in the first place, a loom simply being plonked down in the main room. In the Stroud valley a differential survival of the bigger buildings, with the existence of some documents and associated literature, has channelled students away from rural domestic industry towards excessive concern with the great woollen mills.

The successor industries that came after cloth making and trading in wool are known to specialist historians. But one of the most significant activities, wood-working in sheds, has been overlooked and when it took place out among the trees manufacturing may as well never have happened. A few saw pits may still exist but they will be undateable hollows in the ground. As Chapter 12 demonstrates, even the installation of steam engines has left no mark and I chanced on a clue only when casually examining Bryant's 1824 map of Gloucestershire, followed by searching obscure estate records for mentions of the engines.

The fact that buried landscapes and physical artefacts can be guaranteed to be incomplete is another reason for not relying solely on them. Fairford Park is an example. The stables remain in use as offices nowadays but the house and its top-rate gardens have gone. A fair outline of the parkland in its eighteenth-century form can be made out, despite the superimposition of an American hospital-cum-Polish camp, both expunged in turn as completely as a prehistoric field system might be. Major features like the ha-has surrounding the deer parks do survive and sketch an outline of the layout, but only an outline. Like the vanished wood-working industries north and south of Stroud, the full scene can be conjectured only in the mind's eye. The ground evidence has to be supplemented by documents, among which Johannes Kip's plate of Fairford Park is a fortunate guide. No such elaborate picture is likely to be available for less illustrious sites, especially those in landscapes of production, which mutate more than estates fixated on lavish consumption.

Reconstructing the fully operational landscape of any date is hampered by omissions among all forms of evidence. More difficult still is working out precisely how institutions of the past actually worked, which is a necessary prelude to judging their economic significance and the motivations of the participants. Post hoc theory may get in the way: Garrett Hardin's rather outrageous suppositions concerning the management of common land, now fixed immovably in the minds of social scientists, have become prime examples.[2] They involve assumptions about the behaviour of villagers that fly in the face of common prudence, never mind modern understanding of collective behaviour. Were people really so foolish as to overstock the commons or so careless as to permit their neighbours to do so? Antisocial behaviour can always be found: speaking in 1796 about Otmoor, Oxfordshire, the Bath & West Society complained that wherever parishes shared a common there was no regular stint and each household turned out as many animals as it pleased.[3] But no other localities are identified and in any event intercommoning was on the decline. Population growth had put a premium on grazing rights and they were increasingly regulated. Hardin's assumption that every villager would overstock if he could is unjustified. The mention of stinting gives the game away, because a stint of the animals turned out was the usual device, enforced after recurrent presentments for infringing the village by-laws. The wide distribution of lot meads also shows that constraints on the sharing of resources were the norm.

In the past, decision-making about allocating resources and eligibility for making use of them was typically communal, especially before the enclosure movement or in some of the areas of eastern and western England that lay largely outside its scope. Equity was aimed at, rather than maximising income, profit or growth. Remnant institutions evolved to manage such matters were to be found until recent decades—courts leet, beating the bounds, common field systems, lot meads—and can be taken as indicating how they originally worked, providing care is taken to exclude modern revivals. Normally the former arrangements had been superseded or undermined by individualist arrangements and some have been almost wilfully abandoned or closed down. The one intact common field village, Laxton, Nottinghamshire, is repeatedly threatened with dismemberment by the very national authorities responsible for it. The later stages of many antique practices are sad, and not just in the eyes of sentimentalists, lovers of tradition or admirers of human diversity. Licences for net-fishing, especially for salmon, have been reduced by the authorities and some practitioners have been bought out by rich fly-fishermen whose own catch has not been curtailed. Traditional practices have been squeezed nearly to extinction.[4] They were admittedly vestigial by the twentieth century; all the more reason, it might be thought, to treat them with reverence. The conclusion is that surviving medieval practices of land allocation and resource use are more threatened than is much endangered wildlife—and ironically would be easier to preserve.

The fragility of traditional institutions was exposed by the failure of nature conservation bodies to ensure that the Yarnton lot mead ceremony would continue. Just how insouciant the public authorities are with respect to ancient rights is even more forcefully shown by the example of Groveley Wood, Wiltshire.[5] Local people had always possessed the right of estovers (picking up firewood), despite efforts by the Wilton estate to suppress it. The estate tried again in 1825 and when a nineteen-year-old woman from Barford St. Martin called Grace Reed and three other women from Wishford went to gather wood they were arrested and imprisoned. Amazingly enough for the date and in view of the forces against her, Grace hired a lawyer to contest the action. The courts found against the Wilton Estate and upheld the right of the inhabitants of

Barford and Wishford to collect wood as they wished. This is a heartening story so far but ends in a mixture of farce and tragedy. In 1970 the right of estovers was lost when the parish council simply did not comprehend a document passed on by the Wilton Estate and thus failed to maintain the right. No doubt all this was legal, though it was in fact legalistic. The shocking aspects are that the legal system is so rigid; that so time-honoured a right could be permitted to evaporate overnight; and that no outside body stepped into defend it. The defence of ancient traditions is, to insist on a point already made, more vulnerable to greed and the hide-bound actions of local government than are the rare species on which conservation energies are spent.

Landscapes and the Needs of Conservation

What, then, should we conserve or even revive among the residues of the past? Surviving institutions dedicated to resource allocation have already been discussed: they should be prioritised as equally significant as built monuments. Beyond that, whole areas representative of certain past uses of the land deserve attention. Individual buildings and archaeological sites receive the lion's share of the funds, though 'lion' may be too majestic a term when all historical preservation is underfunded. Compared with more visible artefacts, landscape is an orphan: a running sore in this respect is the neglect of footpaths, overgrown and blocked as many are and left to the chance of local initiative for registering others.[6]

Under the heading of landscapes, habitats should figure. Nature hotspots may be conserved in the form of reserves but the everyday countryside is left to management by farmers and game preservers, subject only to band aid measures of environmental subsidy. It might be hypothesised that the market in the shape of relative prices produces the result that society approves; this would be acceptable were the market not so distorted by gross inequities of ownership, bureaucratic regulation and inconsistent support from the general tax revenues. The late Michael Shrubb, a Sussex farmer and ornithologist, was candid about this. He referred to the 'perfectly respectable' argument that subsidies should be removed and farming left to find its own level. If it shrank into the tracts most attractive to agricultural investment, there would be plenty of opportunity to conserve wildlife in the marginal areas from which cultivation had been withdrawn. His fellow farmers seemed to think that these areas would revert to wilderness, presented as ‘a sort of bogeyman to frighten us’. Shrubb’s riposte was, ‘what’s wrong with that, it’s the whole point’.[7]

Conservation suffers from limited historical vision. The problem is what baseline period is to be aimed at. This is dealt with arbitrarily, as the example of the chalk downs demonstrates. Arbitrariness is revealed in its most acute form by the taste for rewilding, which takes two forms, the first of which is abandoning current land use and letting habitats revert how they will. Too much has changed and continues to change in the face of diverse factors, such as invasions by alien species, to imagine that benign neglect can create anything other than a hodgepodge. It may merely result in interesting scrubland of the type Michael Shrubb seems to have envisaged, which would however have little warrant in any known historical regime. The second meaning involves reintroducing species eliminated over the course of past time, which can only produce a hodgepodge of its own. Rewilding efforts may be costed but the calculations do not include the opportunity cost. The record of nature bureaucracies in land management is dire but the answer can hardly be private initiatives as dubious as one chosen here almost at random—the fad for introducing Australian eucalypts. Under English conditions these are environmentally unfriendly trees and the wood may end up sold for fuelling emission-producing wood burners.[8] New threats emerge from even less responsible people, for instance juniper is said to have been cleared from Sussex and Hampshire because it is smokeless and the owners of wood burners, anxious to evade pollution restrictions, have seized it.

Sensitive and well-thought-out combinations of regulation and incentives will be needed to combat the endless cross-cutting assaults on the countryside and its wildlife. All one can hope is that efforts are better informed than they have been. One of the reasons why environmental history was so late to emerge as a formal subject probably lies in the premature specialisation in English education whereby school pupils are segregated early into either science or arts streams. Interdisciplinary studies would make a better background but, although they are endlessly canvassed, they seldom catch on. They are unattractive to people trained and with their human capital vested in specific subjects which, despite some broadening of university syllabuses, means almost all of us.[9]

Coupled with this, observational skills are neglected, which has reinforced a tendency to study past environments through what archaeologists call desk-based assessments. These involve examining published and archival sources rather than carrying out fieldwork. In practice, archaeologists are quite broad-minded and back up their reports with site inspections when they can. On the contrary, students of more recent periods are not typically encouraged to do fieldwork, so that the inherent weakness of archaeology—lack of documents—is matched by the historian's weakness of restricting him- or herself to the library or record office. A long time ago R. H. Tawney said that what historians need are not more documents but stronger boots and his aphorism used commonly to be set as an exam question ('discuss'). I am referring of course to broad tendencies, to which there are plenty of exceptions, although not enough: the injunction is mostly honoured in the breach.

A practical measure would be to insist that biologists, farmers and land managers do not drop history from their schooling so early, in order to equip them with a better sense of how the countryside changes and has changed over time. Greater historical awareness would increase appreciation of its slow interconnected evolution and the clues this has left on the land. Historians as well as scientists should be encouraged to read outside their field, just as scenario planners are. Not only is fieldwork minimised or missing from those parts of the history syllabus where it would be appropriate, but historians—breathing sighs of relief—quit the sciences, as well as giving up on statistics, early in their teens. Natural history, where it is taught at all, is left to the primary schools, as if it is inherently childish. Scales might fall from more eyes if students were made aware, for instance, of botany.

The landscape carries many botanical indicators of former occupation and exploitation. The example cited in this book is the presence of gorse, almost wholly an anthropogenic species and an unexpected witness to sites of cultivation. A multiplicity of other instances can be found. The best known indicators are crop marks, which were recognised in the first forays of aerial archaeology and are now a standard means of finding sites that are barely visible on the ground. The sites of old buildings are sometimes revealed by clumps of stinging nettles, since nettles are encouraged by residual potassium and nitrogen in the soil. Butterbur may act as an indicator because it was used, and gets its name, from wrapping butter in the large leaves. Garden plants persist on abandoned house sites while gardens which are still cultivated may be approximately dateable by the trees and flowers in them. Later introductions and additions may muddy the waters but some species cluster at distinct periods, such as the monkey puzzle trees which became stylish in the mid-nineteenth century.

Environment and Economy

One of the strengths of economics is to expect to find repetition and similarity among incidents that historians are more inclined to treat as unconnected events. Overgeneralisation is a danger but better that than history as a congeries of separate happenings with no unifying or clarifying themes. This is a splitters-and-lumpers debate, with the historically inclined prone to splitting and economists content with large impersonal categories. Both extremes are unhelpful: splitting atomises the world and puts out of court any explanation beyond haphazard individual motives, while lumping buries motives in undifferentiated masses. The answer is a middle way, which is always hard to find. No sharp-edged criterion assigns behaviour to one category or another and the issue can be settled only by exerting judgement. For that to carry conviction, sensitivity to the preconceptions of all sides is needed. There is plenty of work for educationists!

I do not accept that behaviour is totally random and lacking in pattern. History may not repeat itself but incentives and disincentives can be seen to elicit common types of response. What recurrent features were at work in our (deliberately) disparate set of environmental case studies? One was substitution. At times locations of activity, resources employed and whole industries were altered or replaced. A striking example, when an industry was replaced, was the transition of cloth producers outside the Stroud Valley into becoming suppliers of raw materials, in short moving from manufacturing to trading in wool. This was not so much businesses relocating as changing on the spot like chameleons, as fortunate towns did very profitably. An example of a shift from one raw material input to another was the way in which gunmakers took to using beech wood instead of walnut when supplies of the latter were running out.

Among recurrent patterns was the awkwardly named disintermediation, a term that comes from banking and means cutting out the middleman. The case can be made that this was a frequent feature of economic advance adopted by different occupations at different periods, for instance when the eighteenth-century wool-staplers of Cirencester were left stranded by the new habit of merchants riding around the farms and securing supplies of wool before they were brought to the public market. Later something similar happened to the Wessex sheep fairs, which were short-circuited by dealers buying directly from the farmer and selling the animals on without the palaver and fees associated with fairs. When railways arrived they too engrossed part of the drovers' work, and the flexibility of the lorry entirely replaced the movement of sheep or cattle on the hoof.

A fundamental shift in location and methods was the seventeenth- and eighteenth-century adoption of clover, sainfoin and other fodder crops on the light-soiled hills. This was a disruptive technology, undeniably transformative, putting inexorable pressure on the price of animal feed. It poured new wine into the old bottles of agricultural practices and led to a redistribution of the main zones of arable farming to the hill country, where rotations of grain and livestock could supply both products jointly. In response, the vales ceased to specialise in grain and concentrated more on supplying cattle and dairy products. In one dimension, time, the 'fodder revolution' may seem to have been slow (although that is a modern impression) yet in the spatial dimension it was extensive, rearranging agriculture across the great bands of hill and vale formed by the country's geological structure. As part of the change, there was a fall in the relative value of floated water meadows, which had once been such potent novelties.

Disruptive technologies were part and parcel of economic progress. Nor will substitutions surprise economists, since the profession is disinclined to believe laments, loud at all periods and in many circumstances, that the exhaustion of resources must spell ruin for existing producers and industries. Agreed, no iron law says that alternative resources are certain to be forthcoming or that means will be available to exploit them. Even when new technologies do appear, incumbent producers do not necessarily adopt them. Nor are novelties guaranteed to succeed: the transience of steam manufacturing in the Cotswold woods shows this. It is plain there were winners and losers among the businessmen affected by disruptive technologies, though the experiences of the losers may be concealed by the overall economic growth of the centuries we have examined. Economics is oddly unconcerned with the detail of the distributional consequences reflected in the misfortunes of people whose work and homes were made redundant. The interest is more about labour than bosses. While reallocation was taking place individual businesses suffered, but this is downplayed—from modern Olympian heights—as collateral damage in the course of national advance. Collateral damage is still damage whatever the eventual gain from creative destruction.

Interpreting the landscape changes underpinning these processes is sensitive to which places are chosen. Individual localities may prove untypical, something best avoided by studying clusters of sites. Interesting trends may also be missed through selecting periods too short to allow them to develop. Longer periods than standard historical divisions may be needed to expose the ultimate malleability of economic activity. Features that at first seem fixed may shift and evolve more than anticipated. For example, the edge of cultivation slipped to and fro over long periods, two steps forwards and one back, at least on the more impoverished soils. Seemingly timeless institutions have likewise fluctuated over the years, even going out of existence for a spell only to be revived at later periods. Gaps in the histories of lot meads may be cases in point.

These processes took place against a background of unidirectional changes, of which the most noteworthy were the diffusion of fodder crops and the insertion of closes of grass into the common fields. The eventual inversion of agricultural distributions was immensely significant. At a still deeper level there was a progressive decrease in the relative value of permanent pasture: only relative, it is true, for landowners were always eager to preserve stored-up fertility from the opportunism of outgoing tenants. Extended periods are best for recognising the full force of these changes. And beneath all development lay Southern England's pattern of soils and topography, most evident on standing back to consider groups of counties rather than just one. The ecological underpinning affected production costs, doing so most strongly when technologies were weakest. The locations of manufacturing and farming altered over the generations, when substitutions of factors of production became clearest.

This brings us to two of the great conundrums of environmental history. First, how far were the locations of activity determined by the natural environment, as opposed to institutions or individual enterprise? No unique rule seems to obtain, as is plain when the various distributions of cultivation on heathland are considered. Secondly, how labile were communal systems, once taken as changeless until ruptured by Parliamentary enclosure? Over the long run internal decisions could hollow them out, at least in part. The stickiness of communal decision-making was relaxed. With the growth of the market, social rules were subtly modified, not that evidence is always easy to detect. But a concern for equity in communal farming was not all it seemed. Dispute avoidance may have been more significant, until little hierarchies of wealth arose within apparently egalitarian communities. The more successful or vociferous individuals were then willing and able to subvert the old arrangements.

The role of individuals in historical processes is tricky to evaluate. It is clear that decision-making was the work of specific people whose names are sometimes known to local historians. This is evident, say, among the last men to sustain lot mead ceremonies or the prominent individuals who had invested in building the wool trade's market halls. They can be identified. Might not others have stepped forward if those who actually did so had not appeared? The issue is whether structural forces determine outcomes, as opposed to the strivings of actual people—did blind arrangements of incentives dictate the observed results or should we turn to what was once called a 'great man' theory of history? I have known a sociologist decline to be shown places of interest on the grounds that, had an event not occurred there, it would have happened somewhere else and some other person would have unavoidably taken on a task dictated by the structure of incentives. This reduction of history to the interplay of automata is not credible: would someone with equivalent characteristics have replaced any of the great leaders known to the history books? Hardly. Environmental history was made by the actions of myriad individuals, constrained without doubt by the currents of their times and ordered by the potential of the landscapes they lived in, but in the last resort exercising their own initiative.

At the local level, its vagaries make historical change hard to predict. Circumstances differ but landscape observation and then written sources can help to uncover the thoroughly unexpected. In this book, the most striking example concerns the use of steam engines in the Gloucestershire woods and the role of the distant industry that employed them: making gun-stocks with which Birmingham arms manufacturers could supply the slave trade (Chapter 12). Of the two types of evidence, landscapes are primary. They are not to be taken at face value because they have never remained as stable as they seem, but have been moulded and remoulded by human action. Each transition or shift in location adjusted to and then altered the relative costs of resource exploitation. Landscape evidence can thus be persuaded to throw light on the deep, abstract themes that have recurred throughout the past, for instance disintermediation and disruptive technology. In this way, the most obscure features of the rural scene connect with the whole process of economic development.

Notes

1. An example that does not figure here but appears in other publications is the history of settlement fires, which translated from casual observations of similarities in small town streets to become contributions to much broader disaster studies. See e.g. E. L. Jones, 'The Reduction of Fire Damage in Southern England, 1650–1850', *Post-Mediaeval Archaeology* II (1968), pp. 140–149, or E. L. Jones and L. E. Frost, 'The Fire Gap and the Greater Durability of Nineteenth-Century Cities', *Planning Perspectives* 4 (1989), pp. 333–347.
2. Eric L. Jones, *Locating the Industrial Revolution: Inducement and Response* (Singapore: World Scientific, 2010), pp. 218–219; Eric L. Jones, *Barriers to Growth* (Basingstoke: Palgrave Macmillan, 2020), passim.
3. *Letters and Papers* of the Bath & West Society, VIII (1796), p. 115.
4. Eric L. Jones, *Revealed Biodiversity: An Economic History of the Human Impact* (Singapore: World Scientific, 2014), pp. 237–244.
5. W. H. Hudson, *A Shepherd's Life* (London: J. M. Dent, 1936), pp. 154–155; Simon Reed, *History of Barford St Martin* (Barford St Martin Parish Council website, 2000).
6. This is a further instance of the inadequacy of government. See Eric L. Jones, *Landed Estates and Rural Inequality in English History* (Basingstoke: Palgrave Macmillan, 2018), Chapter 5, Road Capture. Suggestions for action are given by Jack Cornish, 'Five Ways to Save the Ancient Ways: A Challenge to All Local Historians', *Oxfordshire Local History* 10/3 (Summer 2020), pp. 61–64, but the immense difficulties placed in the way are expounded in Nick Hayes, *The Book of Trespass* (London: Bloomsbury, 2020).
7. Michael Shrubb, *Birds, Scythes and Combines: A History of Birds and Agricultural Change* (Cambridge: Cambridge University Press, 2003), p. 328. Shrubb was well aware that abandoning marginal land might have perverse effects.
8. *The Times*, 30 June 2020.
9. 'Less than half of economists surveyed believe that knowledge gained from multiple disciplines was better than that gained in one.' Review by Quinn Slobodian, *Dissent* (Winter 2020).

Index

E. L. Jones, *Landscape History and Rural Society in Southern England*, Palgrave Studies in Economic History,
https://doi.org/10.1007/978-3-030-68616-1

J

K

L

M

N

FSC
www.fsc.org
MIX
Papier | Fördert gute Waldnutzung
FSC® C083411

Kolibri 360 GmbH
Ferdinand-Jühlke-Straße 7
99095 Erfurt, Deutschland
produktsicherheit@kolibri360.de